The

A v… …rvey of British … and edited by John Gilmour, Sir Julian Huxley, Margaret Davies, and Kenneth Mellanby. The New Naturalist has been described by the *Listener* as 'one of the outstanding feats of publishing since the war', and by *The Times Literary Supplement* as 'a series which has set a new standard in natural history books'. Founded in 1945, it now contains more than 50 volumes, of which the following have already appeared in Fontana:

The Snowdonia National Park WILLIAM CONDRY

The Highlands and Islands
F. FRASER DARLING & J. MORTON BOYD

A Natural History of Man in Britain
H. J. FLEURE & M. DAVIES

The Trout W. E. FROST & M. E. BROWN

Wild Flowers J. GILMOUR & M. WALTERS

The Open Sea – Its Natural History
Part I: The World of Plankton ALISTER HARDY

Insect Natural History A. D. IMMS

The Life of the Robin DAVID LACK

Life in Lakes and Rivers
T. T. MACAN & E. B. WORTHINGTON

Climate and the British Scene GORDON MANLEY

Pesticides and Pollution KENNETH MELLANBY

Mountains and Moorlands W. H. PEARSALL

The World of the Soil E. JOHN RUSSELL

Britain's Structure and Scenery L. DUDLEY STAMP

The Sea Shore C. M. YONGE

The Peak District

K. C. Edwards C.B.E. M.A. Ph.D.

assisted by H. H. Swinnerton C.B.E. D.Sc.
and R. H. Hall F.L.S.

COLLINS

The Fontana New Naturalist

First published in the New Naturalist Series 1962
Revised edition first published in Fontana 1973

Typeset by Jolly & Barber Ltd Rugby
Printed in Great Britain by
William Collins Sons & Co. Ltd Glasgow

Contents

Plates

THE PEAK DISTRICT

Maps and diagrams

Reference to a topographical map will be an advantage in perusing this book. The most suitable one for the purpose is the excellent Ordnance Survey One-Inch Tourist Map of the Peak District, published in 1957 and revised in 1963. This sheet, which is on a scale familiar to all ramblers and nature-lovers, shows the boundary of the National Park

Editors' Preface

The situation of the Peak District in the heart of England as an island of varied hill land, often of spectacular scenic charm, almost surrounded by industrial lowland in whose cities and towns, often gloomy and grimy, live a quarter of Britain's population, made it a natural choice for the first of our National Parks. The old geography books not infrequently referred to the Pennines as the 'backbone of England' and here, at their southern end, the earth's bony framework of older rocks appears not only at the surface, but towering to heights whose grandeur belie their modest elevation.

Nottingham is but one of the cities which adjoin the Peak District, but one from which there are natural lines of entry by some of the most charming valleys. Certainly the members of the University of Nottingham have long shown a particular interest in the area – since well before the University College became the University – and it is thus especially appropriate that Professor K. C. Edwards should head a team of his colleagues and friends to act as principal author as well as editor of this composite volume. As Professor of Geography he has spent most of his academic life at Nottingham and, as the pages of that successful journal *The East Midland Geographer* show clearly, he has done much to encourage the scientific study of the surrounding area, including the Peak District. In this he carries on the tradition established by the octogenarian Professor Emeritus of Geology, the author of the volume on *Fossils* in the New Naturalist Library, Professor H. H. Swinnerton, who contributed two of the basic chapters in the present volume.

Four main rock types dominate the Peak District – coarse sandstones or 'grits', shales, massive limestones and, less conspicuous, the old volcanic rocks locally known as toad stones. It so happens that these rock types offer very different resistance to the forces of nature and so The Peak has become a region of sharp contrasts – from the intimate wooded dales to the windswept, boggy moorland heights. The habitats afforded are correspondingly rich and varied and so naturally are both plant and animal life and the response of man as farmer. Mineral wealth has added to the variety of man's responses, so that in addition to its other attractions the Peak District exhibits fascinating fragments of the story of man's occupation of the area from prehistoric times to the present.

A region of such difficult relief has long offered a challenge to man – in selecting sites for his settlements, in finding routes for his roads and railways, in utilising the varied scattered resources. Each of these aspects is taken up in turn and we are led to see clearly the competing claims on Peak District land – for farming, grazing, forestry, water supply, recreations, sport, nature conservation and others – and how the solution may lie in the careful application of the principles of multiple use within the framework of a National Park administration.

Although with the names of K. C. Edwards and H. H. Swinnerton only that of the botanist R. H. Hall appears on the title page, Professor Edwards has been able to incorporate observations by many workers in many fields, and the resulting volume is one which we are confident will have a very wide appeal. As his long association with the Ramblers' Federation, the Youth Hostels Association and many field bodies will show, Professor Edwards believes the way to see and know The Peak is on foot. His book has thus a special appeal to the legions of ramblers who use the National Park every week-end, winter and summer alike.

The Editors

Author's Preface

So many books have been written about Derbyshire and the Peak District that there would seem little excuse for adding to them. The reason for doing so lies in the promotion of a National Park in this area some years ago. This timely and widely-acclaimed decision has given official recognition to The Peak as one of the regions of outstanding scenery in Britain. It has also heightened the interest of those who find enjoyment in the open country and of the many people whose inclination is the study of wild nature. To thoughtful and observant people who require something more than the conventional guidebook, it is felt that an account of the present landscape of The Peak, of how it came into being and of the activities it supports, would be of some value. Moreover, the ground covered by this volume extends beyond the limits of Derbyshire, for the territory embraced by the National Park includes portions of several adjoining counties.

The plan of the book is intended to be simple and logical. The earlier chapters deal with the strictly natural aspects of the area from the story of the rocks to the formation of the present land surface and its vegetation cover. Then Man enters on the scene and the later chapters describe how in different ways Man has imposed, and continues to impose, his cultural imprint upon the setting prepared by Nature. The concluding chapter examines those aspects of The Peak which make it so acceptable to the community as a National Park.

In preparing the book much help has been given by various authorities on the area, which is gratefully acknowledged. Chapters 2 and 3 have been written by H. H. Swinnerton, C.B.E., D.Sc., Emeritus Professor of Geology in the University of Nottingham; Chapter 6, together with parts of Chapters 5 and 7, have been contributed by Mr R. H. Hall, F.L.S., a specialist on the botany of The Peak, while the section on Fungi has been prepared by Dr C. G. C. Chesters, Professor of Botany in the University of Nottingham. The late Mr F. A. Sowter assisted with the section on Lichens. For the remaining chapters, two sources of information have proved invaluable. These are the unpublished thesis, *The Peak District National Park:*

a regional study of an amenity area, by Mr G. J. Mosley, M.A., of the University of Nottingham, and *The Report and Analysis of Survey of the Peak Park Development Plan* by John Foster, A.R.I.B.A., M.P.T.I., the first Planning Officer to the National Park Planning Board. Two local publications were also extensively used, the *Journal of the Derbyshire Archaeological and Natural History Society* (annually) and *The Derbyshire Countryside* (bi-monthly) of the Derbyshire Rural Community Council.

In preparing the Appendices on Bird Life and Fish Life in the Peak District I have been helped by Mr E. L. Jones of the University of Nottingham, Mr C. M. Swaine of the British Ornithologists' Union and by Major J. I. Spicer, M.B.E., Chief Pollution and Fisheries Officer to the Trent River Board. I am most grateful to them, and also to Miss D. A. Clarke, Sub-Librarian at the University of Nottingham, for her kind assistance in assembling material and compiling the index.

Moreover, to the many friends and acquaintances who have generously made available their special knowledge of The Peak and to the numerous countryfolk encountered on the moors and in the dales who unwittingly contributed by their ready and forthright response to questions, grateful thanks are expressed.

K. C. Edwards

PREFACE TO THE SECOND EDITION

Since the first edition a number of changes have occurred in and around the Peak National Park, most of which affect the human occupation rather than the natural environment. Assistance with the revision by Mr J. M. Smith, M.A., F.R.T.P.I., a senior official of the Derbyshire County Planning Department, is gratefully acknowledged. In 1966 Professor Swinnerton died and in 1971 Professor Edwards retired to become Emeritus Professor of Geography in the University of Nottingham, while Mr H. Hall is a member of the Derbyshire Naturalist Trust and is on the Management Committee.

K. C. Edwards

THE PEAK DISTRICT

1. Introductory: The First National Park

> By the side of religion, by the side of science, by the side of poetry and art stands natural beauty . . . the common inspirer and nourisher of them all.
>
> *G. M. Trevelyan, O.M.*

National Parks are tracts of country of outstandingly attractive scenery which are specially protected against adverse change and reserved for public enjoyment. In a country like our own, in which a high proportion of the population is concentrated in large cities and industrial districts, there is a real need for the setting aside of particular areas where townsfolk may find relief from the pressing throng and enjoy open-air recreation amid surroundings which bring them close to nature. In national parks, moreover, the preservation of natural scenery is safeguarded as well as the surviving haunts of wild life. Indeed, for the study of living forms in their natural environment, whether plant, insect or animal, such areas are of special scientific value.

NATIONAL PARKS IN BRITAIN

Although the provision of national parks in Britain is a recent development, the idea of reserving selected areas of our finest landscape for the enjoyment of the public is by no means new. It stems in fact from the ideas connected with social betterment arising from conditions in the nineteenth century. The movement for national parks began well over half a century ago but not until the closing stages of the second world war did it gain official recognition.

In other countries national parks of various kinds have long been in existence, one of the earliest being the famous Yellowstone Park in Wyoming, established in 1872. Like others which followed it in America, this great reserve was 'dedicated and set apart as a public park or pleasuring ground for the benefit and enjoyment of the people', words which expressed the aspirations of many who later advocated similar projects in our own country. In a long-settled, densely populated land like Britain, however, with much of its surface under private ownership, the problem of public access to areas of scenic attraction has provided a formidable obstacle to the realisation of such hopes. The contrast between the old and new countries in this respect was clearly demonstrated in 1885 by two events. In that year, within the space of a few months, James Bryce's Bill to give access to mountains in Scotland was rejected by Parliament at home, yet a proposal to create the Banff National Park in the Rockies was accepted by the Government of Canada without opposition. But Bryce's attempt marked the beginning of a long campaign for the acceptance of the national park idea. Apart, however, from an inquiry made by the Addison Committee in 1931 and the passing of the Access to Mountains Act in 1939, the provisions of which were made inoperative by the second world war, little progress was made until recent years.

Meanwhile voluntary bodies had given active support to the movement and the effect on public opinion contributed much towards ultimate success. In 1935 two of the leading organisations, the Council for the Preservation of Rural England and the similar body for Wales, set up a Standing Committee for National Parks, which included representatives of many supporting interests. The Standing Committee worked unremittingly and proved a valuable instrument for educating the public and arousing interest among widely different sections of the community.

During the war, when much thought was devoted to post-war planning and reconstruction, the case for national parks was repeatedly stressed. Official publications such as the Report of the Scott Committee on Land Utilisation in Rural Areas (1942) and the White Paper on Control of Land Use (1944) gave firm support to their formation, while the Report on National Parks (1945) prepared by John Dower for the Minister of Town and Country Planning, by its cogent argument and its deep personal conviction, at last set the course for realisation. The Dower Report was followed by a detailed investigation resulting in the Hobhouse Report on National Parks (1947), of which the chief recommendations formed the basis of the National Parks and Access to the Countryside Act passed two years later. Indeed, so favourable had become the climate of opinion that the Act was passed without a single vote against it in either House.

Briefly, the Act of 1949 provided for the setting up of a National Parks Commission which is responsible for the designation of each individual park in England and Wales and for advising the particular authority chosen to administer it. The park authority, consisting largely of representatives from the existing local authorities within whose areas the park falls, undertakes the management and planning of the park. Other provisions of the Act dealt with important issues affecting public rights of way, access to open country and wild life protection. A further measure (under section 87) permitted the designation of smaller areas of outstanding natural beauty, not as national parks but as areas requiring the assistance of the Act in the interests of nature preservation and the protection of wild life. Several of these have already been designated, such as the Gower Peninsula and the Quantock Hills, and it is likely that many more will be added.

As a result of the Act, ten national parks have so far been established, of which the Peak District was the first. The others are the Lake District, Snowdonia, Dartmoor, Exmoor, the Pembrokeshire Coast, Brecon Beacons, the North Yorkshire Moors, the Yorkshire Dales and Northumberland. Varying greatly in character and extent, these form a system having an aggregate area of nearly 4,750 square miles, or one-twelfth of the total area of England and Wales. This is surely an achievement which meets the long-felt needs of our highly urbanised people and which at the same time will secure large tracts of unspoilt country from inappropriate and unwarrantable forms of exploitation. These conditions were established more precisely by the Countryside Act, 1968.

In Scotland the situation is different. The National Parks Act of 1949 referred only to England and Wales and no equivalent legislation has been passed for Scotland, except that the section of the Act dealing with nature conservation was made applicable to the whole of Britain. The Forestry Commission has nevertheless established a few national forest parks in Scotland, and although these have not the full status of national parks, they offer to the public increased facilities for access and open-air enjoyment.

The national parks of Britain differ from those of most other countries. To begin with, there are no large areas of wild scenery available for preservation like the huge National Parks of the U.S.A. or the great game reserves of Africa. Instead, the areas designated as parks are all of comparatively small extent and all are inhabited, even though the density of population in most cases is quite low. Again, the greater part of each national park is devoted to some form of economic activity such as farming, forestry and quarrying, thus creating problems of public access which seldom arise abroad. In every case in Britain the functions of a national park are added to an established pattern of economic and social life and must operate in such a way so as not to cause serious interference with existing activities. This circumstance naturally calls for much delicate negotiation in co-ordinating the various interests represented within a national park.

There is another important difference between our national parks and those abroad. Whereas the latter usually embrace truly natural scenery unmodified by human action, in Britain hardly any such areas remain. The land has been exploited by Man for so long that the results of his activities have profoundly altered the appearance of the surface. Even the natural vegetation has been greatly modified by various agricultural practices as well as by the chance or deliberate introduction of non-indigenous species. Forests have been entirely removed from some areas and have been created in others, while mining and quarrying have left their scars and debris on land once unspoiled.

In the sense that even our finest scenery, whether mountain, moorland or sea-coast, is in part the product of our cultural history, the national parks of Britain are of a rather special kind. In many respects the effects of long-continued human occupation signify a gain rather than a loss, for these effects tend to heighten the distinctive character of each desiignated area. Each national park bears the impress of its cultural history and this, taken in relation to the physical conditions gives it a high degree of regional individuality which it would not otherwise possess. It is for this reason that the national parks of Britain offer such impressive contrasts in landscape.

THE PEAK NATIONAL PARK

The upland region traditionally known as the Peak District, which forms part of the southern Pennines, has special claims to rank among the earliest of the national parks established in Britain. It was in fact the first to be designated, in December, 1950, although the Lake District and Snowdonia were included shortly afterwards. On grounds of the quality of its scenery alone its claim was irrefutable. The vast open moorlands which spread out from the massive summit of Kinderscout, the green dales of Dovedale and the Manifold Valley, the rocks and caves like those around Castleton and the glistening trout streams such as the Dove and the Wye, have long been enjoyed by large numbers of visitors from all parts of the country. Above all, The Peak has been cherished by the thousands of ordinary people, young and old, from the industrial towns of the Midlands and the North, who find recreation and adventure in the attractions it offers.

Like the National Parks movement as a whole, the promotion of such a park in the Peak District owes much in the first instance to voluntary bodies. Among these the Sheffield and Peak District Branch of the C.P.R.E. has been particularly active. More than any other body it was responsible for the idea in the first place, and from 1939 it has undertaken a great deal of pioneer work, including the preparation of a preliminary map of the boundaries. No less than ten thousand copies of the booklet entitled *The Peak District a National Park*, published in 1944, have been sold. Today, even though the Peak Park is now an accomplished fact, continued vigilance on the part of such bodies is perhaps more than ever necessary. An illustration of this is seen in the recent proposal to build a motor road along the Manifold Valley, a scheme which was happily defeated, again largely through the protests of the Sheffield C.P.R.E. supported by many other like-minded persons and societies.

From the standpoint of its geographical position the claim of The Peak to become a national park was particularly strong, for nearly half the population of England live within 50 miles of its boundary. Not only do the large cities of Manchester and Sheffield virtually adjoin it, but many other industrial centres of South Lancashire, the West Riding of Yorkshire, Nottinghamshire, Derbyshire and the Potteries are only a short distance away. Less than 60 miles away are Liverpool and the rest of Merseyside, while Birmingham and the Black Country, Leicester, Hull and Teesside are all within 75 miles. Moreover, of all the national parks so far created, The Peak is the nearest to London and the most accessible from it by rail or road. The proximity to the park of such a large population has an important bearing upon its use by the public and hence upon problems of management. It can be approached from all directions and is frequented, at all seasons of the year, especially by people from the nearby industrial centres.

Popular claims to The Peak as an amenity area are further supported by the variety of interest which it offers to the more serious observer such as the naturalist, geologist, geographer, and archaeologist. The Peak is a part of highland Britain, yet is readily accessible from the lowland zone. As such, in its natural (including biological) features and its cultural forms, it exhibits elements of both environments. Its distinctive physical composition and its rich cultural legacy combine to form a regional complex with a character entirely its own. To both the week-end rambler and the holiday-maker with an inquiring mind the area affords a rewarding field for observation and study in the open air. To the naturalist in particular it is significant for its examples of true mountain and northern moorland habitats.

The attractions of The Peak have long been recognised by travellers from other parts. They gave rise to a literary cult rather similar to that inspired by the Lake District, though it may well have begun at an earlier date. In 1636 Thomas Hobbes, the philosopher and author of *Leviathan*, published his poem *De Mirabilibus Pecci* (Concerning the Wonders of The Peak), describing in ponderous hexameters the outstanding features of Derbyshire. Provided with an English translation in similar verse form, the poem became widely known and was reprinted several times in the next fifty years. Izaak Walton, whose delightful treatise *The Compleat Angler* (1653) offered pleasant instruction in the art of fly-fishing, shared with Charles Cotton of Beresford, with whom he pursued this quiet sport, a lasting affection for the Dove, the Wye and other Derbyshire streams. Though not accessible to visitors, the Fishing House overlooking the Dove, which Cotton built in 1674 to serve as an idyllic retreat, still stands. With their monograms inscribed over the door it bears witness to the long friendship between these two men. Cotton, besides being devoted to country pursuits, was influential both locally and in London and drew the attention of numerous friends to the scenic attractions of the Derbyshire-Staffordshire border, and in 1681 issued *The Wonders of The Peak*, a eulogistic poem much in the vein of Hobbes. Really in the nature of a guide, its true object is perhaps revealed by the fact that it was published in Nottingham and sold by the booksellers of York, Sheffield, Chesterfield, Mansfield, Derby and Newark, all of them places from which visitors might be expected to start their journeys into the region. Later on Celia Fiennes, a shrewd and observant traveller, describes in her *Northern Journey*, made in 1697, the route through Derbyshire, visiting each of the seven wonders in turn as if to do so were already the conventional tour. To the eighteenth-century writers, who almost invariably depicted the notable features of the area in exaggerated terms, six of the seven acknowledged wonders were natural features while the other was the Duke of Devonshire's great mansion at Chatsworth. Its accepted place in the list is probably due to its inclusion by Hobbes as a compliment to his patron. Not everyone at this time held such favourable views, for Daniel Defoe in his *Tour through Great Britain* (1778) denounced Derbyshire as 'a howling wilderness'. It was left to Edward Rhodes in the early nineteenth century, however, to establish for The Peak a lasting reputation as an area of beautiful scenery with many attractions for the tourist. His book on *Peak Scenery* (four volumes: 1818-23), splendidly illustrated by the artist and sculptor Sir F. L. Chantrey, R.A., who was a native of Derbyshire, was widely read and thus helped to make the area known to people from more distant parts of the country. The discovery of The Peak by visitors from outside was soon to be facilitated by the early railways. The new mode of travel brought the lovely scenery within reach of people belonging to all sections of the community, just as a century or more previously the fashionable spa at Buxton had attracted the well-to-do.

Recently, work by Mr R. W. V. Elliott has revealed the existence of a literary association of a very different kind. By relating descriptions given in the text to actual topographical features, Mr Elliott has shown that in all probability the Staffordshire portion of The Peak between Leek and Macclesfield, which now falls within the National Park, provided the setting for much of the famous medieval poem *Sir Gawain and the Green Knight*. The site of the castle of the Green Knight, to which Sir Gawain came, can be identified as that occupied by Swythamley Hall. Though the castle was evidently fictitious, Swythamley itself was certainly a hunting lodge in medieval times and came to be part of the endowment of the abbey of St Mary and St Benedict of Dieulacres near Leek, with which the origin of the poem may be connected. The Green Knight's hunting grounds can be traced across the Roaches towards Flash and northwards beyond the headstreams of the Dane. The Green Chapel sought by Sir Gawain is doubtless the curious rock-chamber known to Dr R. Plot, the seventeenth-century Staffordshire historian, as Lud's Church and still named as suchh on present-day maps. (There is a tradition that the chapel served as a refuge for Lollards.) From the evidence it is clear that the author of *Sir Gawain* was not only minutely acquainted with this district but possessed a remarkable eye for detail and an exceptional capacity for precise description. It is to be hoped that further research will lead to a closer knowledge of the

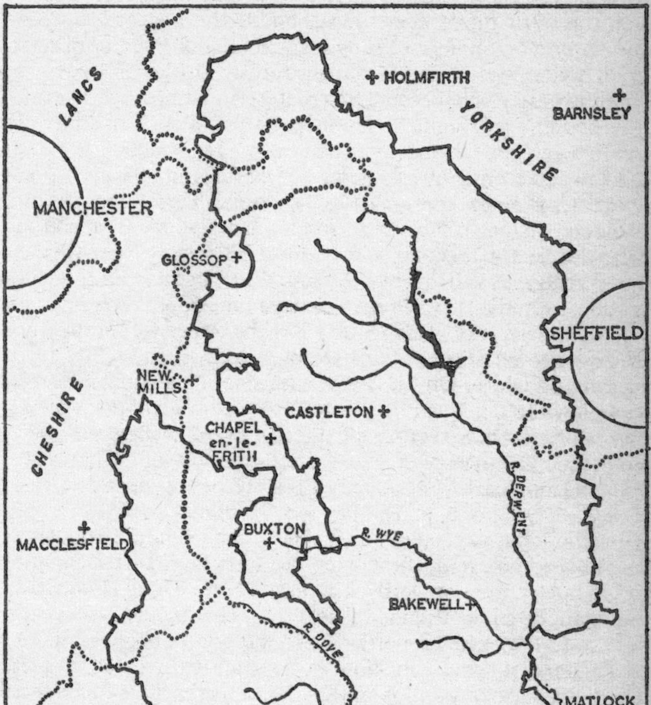

Fig. 1 The boundary of the Peak National Park

unknown poet of the fourteenth century whose work surely ranks with that of Chaucer.

In size The Peak National Park, covering 542 square miles, is not so large as those of the Lake District and Snowdonia, though it is larger than any of the others. It occupies a considerable part of the county of Derbyshire, together with adjoining portions of Staffordshire and Cheshire to the west and the West Riding of Yorkshire, including a small area of the city of Sheffield, to the north (Fig. 1). Its greatest length from north to south is nearly 40 miles and its greatest breadth about 24 miles. The boundary which delimits the park would enclose a broadly oval shape but for the long narrow wedge reaching far into the interior from New Mills on the western margin to a point about five miles south-east of Buxton. This territory was excluded on the grounds of its predominantly industrial character; in it lie the towns of New Mills, Whaley Bridge, Chapel-en-le-Frith and Buxton, while around the last-named the landscape is seriously marred by intensive limestone quarrying and related lime-works. This feature has the effect within the Park of severing the High Peak of Derbyshire from the hill country of East Cheshire. On the south-east side for a similar reason, though without giving rise to a pronounced wedge, the Matlock and Darley Dale section of the Derwent valley was also excluded. On the south and west the towns of Wirksworth, Ashbourne, Leek and Macclesfield have all been omitted, though they lie only a little beyond the boundary.

Incidentally, two points concerning nomenclature should be mentioned here. In the first place, the term Peak District applies to the upland area of Derbyshire as a whole. There is no single mountain or summit named The Peak. The highest part of the area is in the extreme north where the two flat-topped moors of Kinderscout and Bleaklow both reach to over 2,000 feet. The highest point of all is on Kinderscout reaching 2,088 feet. Secondly, a distinction is sometimes made between the northern and southern parts of the upland. These are known respectively as the High Peak and Low Peak but the distinction is a vague one and is not based on altitude alone.

From the standpoint of administration the Peak National Park is managed by a central authority in the form of a Joint Planning Board consisting of 27 representatives, of whom 18 are appointed by the constituent county authorities and the county borough of Sheffield, and the remaining nine by the Minister of Housing and Local Government, now Secretary of State for the Environment. The Peak Planning Board has its own technical department at Bakewell under the direction of a Planning Officer. This form of administration is the only example of the method originally envisaged for the national parks although in one or two other instances the form adopted approximates to it. On the whole in the case of The Peak it has worked successfully. Certainly, by excluding a number of urban centres situated on the fringe of the Park, general planning problems connected with the location of industry and urban growth and re-development have been considerably reduced, enabling the Planning Board to devote its attention more wholeheartedly to the special interests of the Park.

REFERENCES

Pilkington, J. *A View of the Present State of Derbyshire* (2 vols). (1789)
Rhodes, E. *Peak Scenery, or the Derbyshire Tourist.* (1818)
Glover, S. *The History, Gazetteer and Directory of the County of Derby.* (1829-33)
Dower, J. *Report on National Parks in England and Wales.* H.M.S.O. (1945)
Report of the National Parks Committee. H.M.S.O. (1947)
Report of the National Parks Commission. H.M.S.O. (annually since 1950)
Monkhouse, P. J. Some National Park Problems. *Journal of the Town Planning Inst.: 43* (March, 1957)
Porteous, C. *Portrait of Peakland.* Robert Hale (1963)
Poucher, W. A. *The Peak and the Pennines.* Constable (1966)
National Park Guide No. 3: Peak District. H.M.S.O. Second edition (1971)

2. The Rocks and their History

The hills are shadows, and they flow
From form to form, and nothing stands;
They melt like mist, the solid lands,
Like clouds they shape themselves and go.
Alfred Tennyson: In Memoriam

A wanderer returning to his native village in the Peak District finds his favourite haunts unchanged. The dales and streams, cliffs, hills and moors are all just the same as they were in his youth. If, however, he has studied the rocks they tell him that this peaceful, enchanting scene is but an episode in a long and eventful story which moves so slowly that for the brief period of his lifetime it seems to have been at a standstill. His going and coming have been no more than the flicks of a fly's wings. That story is recorded for him in the rocks of the district; in the limestone of the uplands on the south; in the grits capping the moorlands of the north and forming the ridges which girt the uplands on either side; and in the shales which underlie the fertile vales that lie between the areas occupied by these two types of rock (Fig. 2).

THE FOUNDATION ROCK

The limestone teems with fossils which may be seen and collected wherever the surface of the rock has been washed by the rain for a long time or etched by weak acids seeping down from the covering of soil. These fossils are the remains of creatures that lived in an ancient sea which 280 millions of years ago occupied the whole district. How different was the outlook then. Blue sea extended to the horizon in all directions except the south where, in the offing, stood the miniature mountainous island of Charnwood which lay across the area now known as the Midlands. Its rivers were small and carried very little sand and silt into the sea, the waters of which were in consequence clean and clear. The sea was of no great depth and the scenery of its sunlit floor varied from place to place. Here and there were forests of stone lilies, animals that were allied to the starfishes. Each one grew upon a tall stalk built up of rings of lime piled one upon another to a height of eight or ten feet. The main body of the creature was at the top and carried five branching arms spread out like the fronds of a palm tree, to catch both the sunshine and the small organisms upon which it fed. When the stone lily died its flesh decayed and the fairy bead-like rings of the stalk and the limy framework of the arms and body fell to the floor of the forest and in the course of many generations built up deep deposits of calcareous debris.

Out in the open, beyond the bounds of the stone lily forests, lay coral reefs. These were produced by the combined activities of myriads of polyps. Superficially they resembled modern reefs but the structural details of the individual corals were strikingly different. Surrounding the forests and reefs were spacious wastes of mud formed from the shells and bodies of minute organisms which fell in a perpetual drizzle from the waters overhead. Burrowing in the mud or crawling over its surface were many worms and other creatures that fed upon the mud or caught the drizzle as it fell. These latter included lamp-shells, a type of animal that is scarce today but was then varied and numerous and played a much more important part in the economy of the sea floor. Like cockles and mussels their bodies were also enclosed in shells, often prettily shaped and ornamented. They were usually small shells but some were giants a foot or more in diameter.

Of special interest were certain curious molluscs belonging to the far-off ancestral stock of the Pearly Nautilus which lives today in the waters of the Pacific Ocean. Like this, they had shells which were divided into a succession of chambers separated by thin partitions. One large type is known as *Orthoceras* because its shell was straight and not closely coiled like that of the nautilus. Provided with a battery of tentacles round its mouth and with an apparatus for jet propulsion, it preyed upon fishes and other more peaceful creatures. The Goniatites were much smaller and their shells were closely coiled. In them the partitions between the chambers were folded, sometimes in sharp angles (gonia = angle) which suggested the name. These beautiful little creatures may be pictured as spending their days flitting or crawling over the coral reefs and browsing upon the coral polyps.

The shells of all these animals added their quota to the deposits that were being laid down upon the sea floor. Nevertheless,

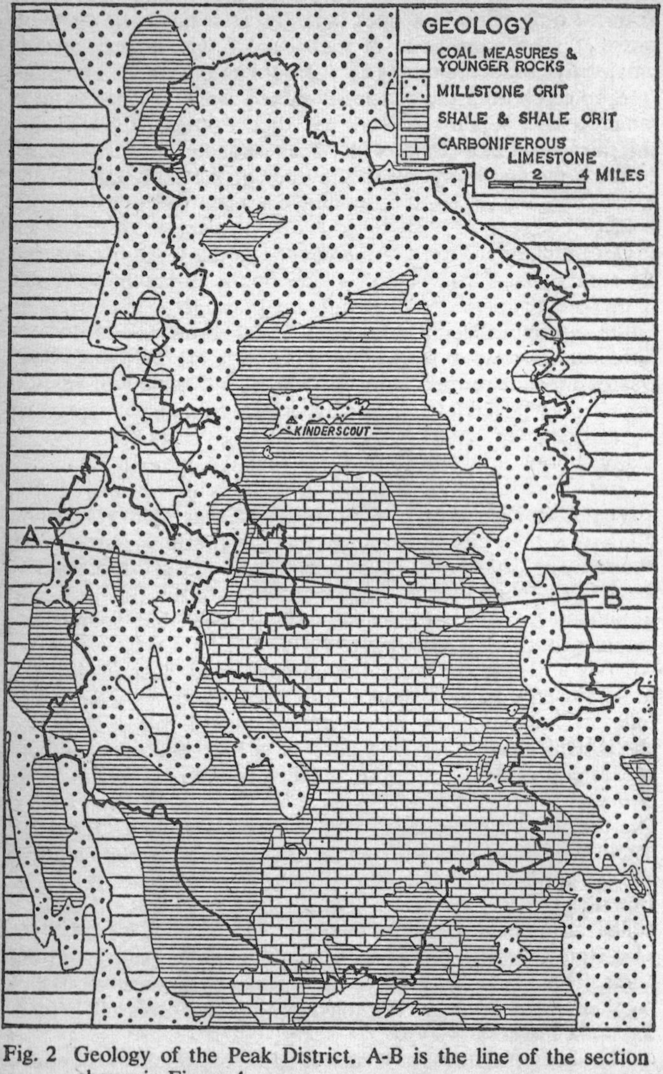

Fig. 2 Geology of the Peak District. A-B is the line of the section shown in Figure 4

though the sea was shallow, it did not become filled, for its foundations were subsiding at about the same rate as the deposits were accumulating. Thus it came about that they ultimately attained a thickness of nearly 2,000 feet.

A temporary but fascinating feature in this submarine scenery was the occasional presence of small volcanoes. The ashes which they shot forth into the waters above settled down and became mixed with the mud beneath. Sheets of lava, full of steam bubbles, were poured out and flowed far and wide over the sea floor (Fig. 3). The dark-coloured rock into which these lavas solidified is known as Toadstone and has a striking appearance due to the fact that the bubbles have been filled with a white mineral.

All the deposits described above consolidated and became the limestone which forms the Derbyshire upland. It is sometimes called the Mountain Limestone but to geologists it is known as the Carboniferous Limestone.

THE LATER ROCKS

The northern shores of the Carboniferous Limestone sea lay 200 miles away and stretched across the centre of the Scottish region. Scotland, at the time, was part of the great North Atlantic continent drained by large rivers flowing southwards. The general geographical picture thus presented was not unlike that of the United States with the Mississippi flowing into the Gulf of Mexico. These rivers carried the debris formed by the destruction of the uplands and by the rain washing the plains into the sea. Deltas of grit and sand were formed and banks built out along the shoreline. The fine muds were, however, carried farther afield and eventually reached the Peak District. Then for some time the sea-water was alternately clear and turbid, but eventually the latter condition prevailed. The mud accumulated and in course of time became those rocks known as the Edale shales which underlie the peaceful meadows of Edale and Darley Dale.

The fauna in the waters underwent a corresponding change. The stone lilies, corals and many of the brachiopods and molluscs departed from the area. A few of the last remained and were joined by other kinds of goniatites and bivalves. Meanwhile the deltas and sandbanks extended and began to invade the district from time to time. The quality of the water also changed from being saline through brackish to fresh. Marine animals

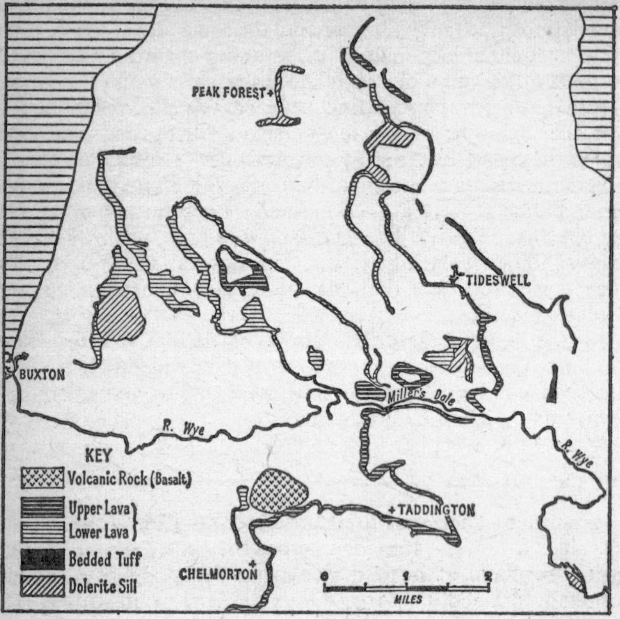

Fig. 3 Volcanic rocks occurring in part of the limestone of the Peak District. (Based on H. H. Bemrose)

disappeared and were succeeded by a less varied and sparse population. At first this included *Lingula*, a curious tongue-shaped brachiopod which had already existed for 250 millions of years from Cambrian times onwards, and was destined to continue in the world for a similar stretch of time until the present day. As the waters freshened still more, *Lingula* and its associates migrated elsewhere and were replaced by crowds of bivalves such as *Carbonicola* which resembled the mussel of the present-day rivers and canals. Towards the close of this phase in the story of the Peak District the occasional influxes of coarse sediments became more copious. Banks of grit and gravel were formed and ultimately became that massive hard rock known as Millstone Grit.

Owing to slight oscillatory movements in the level of the region these deposits were sometimes raised above water level and produced a low-lying landscape of sandbank and water channels. Spores wafted by the breezes, seeds carried by the streams from the continent enabled plants of the northern continent to settle on this new land surface. In the warm moist atmosphere they quickly germinated and produced a jungle growth of fern-like plants and strange-looking trees. Of the latter the smaller ones resembled the Tree Ferns which now grow in the tropical forests of the East Indies. Others, known as Calamites, were closely related to the horse-tails, those tall weeds which look like miniature Christmas trees and today grow profusely in wet waste places. Some of the trees towered to a height of 60 or 80 feet and had trunks as much as five feet thick. The bark was often decorated with scale-like markings which suggested the name *Lepidodendron* (lepido = scale) for these trees. Their branches and twigs had a furry covering of small lancet-shaped leaves. The modern relatives of these trees are not to be sought for in luxuriant forests but on bleak moorlands where the Stags Horn Moss *(Lycopodium)* is to be found straggling through the grass. These plants do not grow from seeds but, like ferns, they reproduce by means of minute pollen-like spores.

Among the undergrowth of ferns and in the pools and sluggish streams were lowly types of four-legged animals, represented today by newts or salamanders, creatures which resembled fishes in that they lay their eggs in water and their young must spend at least the early part of life breathing by means of gills.

These conditions, so different from the present, lasted long enough for deep deposits of vegetable debris to accumulate and become peat. A gentle down-sinking of the whole region then ensued. The sandbanks with their cloak of peat were submerged and were gradually buried under thick beds of mud and clay. The peat, squeezed by the pressure of the increasing load and changed by complex chemical reactions ultimately turned into coal. The mud also was compacted into shale.

The sequence of events last portrayed was repeated a number of times and thus was built up a series of massive grit layers interspersed with shales and occasional seams of coal. These rocks were all destined, in the fullness of time, to play a large part in the scenery and other amenities of the Peak District. The layers of grit and shale, which had attained a thickness of a thousand feet, extended far beyond the bounds of the district and covered a vast area including the north of England and southern Scotland.

The transport of so much sediment from the North Atlantic

continent implies a corresponding destruction of the rocks in its uplands and a general wearing down of its whole surface to lower levels. As one outcome of all this the formation of coarse grit ceased and henceforth only fine sand, silt and mud were transported into the southern waters. Deposition and sinking went on in unison continuously, except for an occasional upward oscillation which converted much of the region, including our district, into an extensive fenland of mud-flats, upon which forests grew once more and thick peats accumulated. When the sinking movement was renewed, sea-water sometimes flowed in and spread everywhere, bringing with it marine animals, especially goniatites. Such marine conditions lasted only for a short time and gave place to a long period of fresh-water conditions when mud accumulated and buried the peat, which was in turn converted into coal.

In this way 4,000 feet of clays, shales and fine sandstone with occasional coal-seams and marine bands accumulated. These are spoken of collectively as the Coal Measures. The economic importance of the coal suggested the word Carboniferous as the most suitable name for the whole sequence of rocks hitherto dealt with, including Mountain Limestone, Edale Shales, Millstone Grits and Coal Measures; and for the long period of time, amounting to 60 millions of years, which they represent.

The countryside now occupied by these Coal Measures is comparatively low-lying. Apart from tip heaps it has a scenic beauty of its own but makes no contribution to the amenities of the Peak District. It does however provide the basis for the livelihood of dense populations. From its cities at week-ends and holiday times streams of hikers and pleasure-seekers pour forth to find renewal of energy and refreshment of mind and spirit in the dales and on the moors.

THE ARRANGEMENT OF THE ROCKS

The process of forming those rocks which make up the Peak District was completed millions of years ago. Two sets of agencies then came into action and recorded their activities in quite different characters. One, which was concerned with rearranging the rocks, completed its work in a relatively short period of time. The other, concerned with destroying and carving the rock of the district, experienced a long interruption of its activities during the Mesozoic Era.

The hiker on Kinderscout or Bleaklow rejoices in the fresh breezes and in fact that he is 2,000 feet or more above the sea. Though he takes that fact for granted it is nevertheless full of significance. Kinderscout is a little plateau with a relatively flat surface defined along its margins by rough, often precipitous slopes of Millstone Grit, for that is the rock which immediately underlies the plateau. When it was formed, that slab of grit was part of an extensive sandbank lying close to sea-level. As already seen, it was subsequently buried under Coal Measures down to a depth of some 5,000 feet, that is to say 7,000 feet below its present level. How then did it come to be at its present height? This was brought about by the joint action of two types of movement of the earth's crust, of which the first must now be discussed.

Just as the process of burial described above was being completed, great revolutionary events began to take place across the south of the British area and the north of France. Mighty pressures in the earth's crust acting from north and south crumpled the rocks along a belt of country about 200 miles wide and heaved them up into a mountain range of Alpine and even of Himalayan proportions. A small backwash of these great events was felt in the region of the Peak District. That threw the Carboniferous rocks into a number of folds which have exerted an important influence upon the physical features as seen today. As one result, the western margin of the district was crumpled into a series of narrow folds which merged into the main axis of the Pennines. Chief among these was a broad fold, elongated from south to north, which is known as the Derbyshire Dome. In cross-section it is asymmetrical and has the form of a wave that is just about to break; that is to say its eastern side or limb rises slowly to a crest from which the western limb drops down more rapidly and is broken by a series of rock fractures or lines of faulting. The slowly rising eastern limb is itself crossed by a succession of minor folds which run at right angles to the crest and which curve southwards beyond the east margin of the dome.

It is, of course, impossible to see a complete section across this and the other folds at any one point, but the visitor wandering from place to place, along footpath, road or rail, will see the rock layers exposed to view in valley sides, in railway cuttings and quarries, or in cliff faces. Sometimes they are seen to lie flat and horizontal. At other times they dip gently or even steeply. Geologists have carefully measured these angles and recorded them upon maps from which they have been able to piece together the structure of the region as a whole (Fig. 4).

The contemplation of such folded rocks creates the impression of a time of turmoil and grievous upset, a time when nature took the rocks in her hands and squeezed them as a child squeezes its plasticine. That impression is, however, quite false. The moulding of the rocks into these varied and complex forms was spread over a period of several millions of years and took place so slowly that except for an occasional earthquake the placidity of the scene was never disturbed.

It must not be supposed that any part of the district ever attained such lofty heights as 7,000 feet or more, for as the crests of the fold rose above the general level of the surrounding country, they became the targets for the destructive action of such agencies as frost, rain and running water. In this way the Coal Measures, while they were being slowly uplifted, were simultaneously skimmed off the upfolds and the debris was transported to the lowlands which during the succeeding or Permian period, were to some extent covered by sea. At the base of the Permian rocks today there is a layer of this debris, a rock called Breccia made up of angular fragments of Carboniferous rocks. This lies unconformably upon the upturned and bevelled edges of Coal Measure shales, sandstones and coal-seams. This unconformity is a silent witness to the fact that when this Breccia was being laid down the Coal Measures covering the upfolds in the Peak District had been removed. The Millstone Grits thus exposed were then attacked by the same destructive agencies. Being more resistant they were not so completely destroyed except over the south of the dome where the limestone core was exposed to view. Along its margins the limestone layers tilted downwards and disappeared under the surrounding shales and grits.

In this way the general surface plan of the district was established no less than 220 million years ago. In the southern half it exhibits a large roughly oval area of limestone encircled by Millstone Grit and shales with grit layers, which are arranged in a pericline; that is to say they dip outwards from the centre of the limestone dome. In the northern half of the area the grit covering remains. There the rocks are horizontal or nearly so along a north-south axis, but on either side of this they dip outwards beneath the east and west flanks of the dome.

Meanwhile, the mighty folding movements along the axes

of the Armorican mountains had ceased but the destructive agencies went on working. As long as any part of this intensely folded zone stood up above the level of the lowlands the rivers flowed down the slopes and spread their burden of gravel, sand and mud across the plains. Thus as the mountains were being laid low the plains were being levelled up. The Peak District with its folds thus became only a small feature in the vast expanse and was therefore gradually covered up and buried under the waste from the distant mountains and remained out of sight throughout the whole of the Mesozoic Era.

A PROLONGED BURIAL

Subsequent to this burial, a cavalcade of great events swept across the Peak District without leaving any trace of its passage upon the present landscape. Nevertheless the story would be incomplete without some reference to those events.

The uprising of the Armorican mountains athwart the path of the dominant rain-bearing winds greatly influenced the climate of the British area. The warm moist climate which encouraged the growth of the Coal Measure forests and swamps passed away, a dry arid climate set in and barren deserts replaced the luxuriant forests and tangled swamps. The Peak District area lay hidden away under the centre of a vast plain swept by hot winds and dust-storms.

Far away on the distant mountains, during rainy seasons, the rivers became swollen into flood. Their waters spread on the plains as shallow transitory lakes and deposited their sand and mud in thin sheets everywhere. Sometimes the flood-waters found their way into and replenished the more permanent lakes. When the dry season set in and the floods ceased the temporary lakes were dried up and the permanent were reduced by intense evaporation. In both cases the salts that were in solution were left behind. The lakes consequently became more and more salty until such minerals as calcium sulphate and sodium chloride were precipitated. In modern times these chemical deposits have been exploited for gypsum and alabaster in south Derbyshire and for salt in Cheshire.

At last this arid period known as the Triassic drew to a close. Its plains began to subside, were carried down below sea-level so that these too became submerged. This event ushered in the second division of the Mesozoic Era, known as the Jurassic Period. At first this new sea tended to be muddy but later

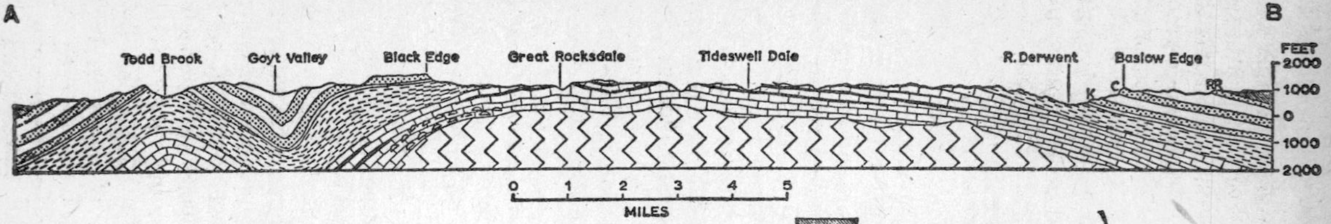

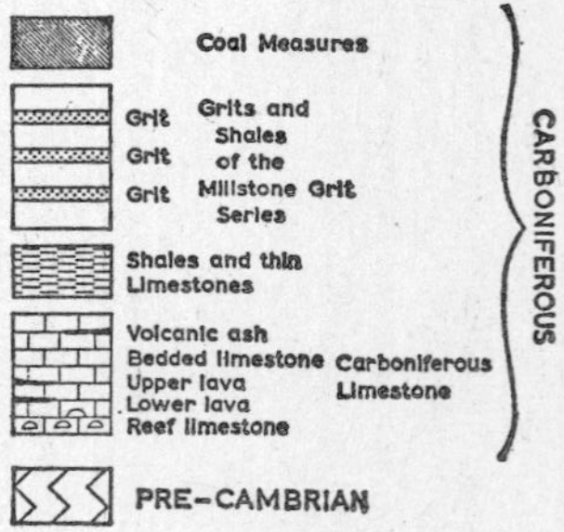

Fig. 4 Geological section across the central Peak District. The section is drawn along the line A-B as shown on the geological map in Figure 2. The individual grits of the Millstone Grit series, being lenticular in form, cannot be traced continuously across the area. Those outcropping in the east are indicated by initial letters as follows: K – Kinderscout Grit, C – Chatsworth Grit, RR – Rough Rock. On the west side Black Edge is formed by the Chatsworth Grit. (Prepared by Prof. W. B. R. King, based by permission on maps of the Geological Survey and other sources)

Fig. 4. Geological section across the central Peak District

its waters were often clear during long stretches of time and then limy deposits were laid down. The latter contained numerous small round pellets resembling the roe of fishes. For this reason the rocks formed from these deposits are known as the Oolitic Limestones (oos = egg).

The animals that came along with these marine waters were many of them similar to but not identical with those that lived in the Carboniferous sea. There were some near relatives of the Pearly Nautilus but they were few in number and all closely coiled. Descendants of the goniatites abounded but their shells were more prettily ornamented and the partitions between their chambers were complexly folded and even frilled. They are therefore distinguished under the name Ammonites. When the waters were clean and clear, stone lilies and corals became common but differed from their Carboniferous allies in many details. Sea-urchins, sea-snails and bivalves also contributed their quota to the building up of the Jurassic deposits and rocks.

Shoals of fishes swarmed in the open waters. Their bodies were clothed in an armour-like mosaic of thick bony scales covered with shiny enamel. They were preyed upon by a new type of animal unknown in Carboniferous waters. These were large reptiles shaped like fishes and almost as perfectly adapted for a life spent wholly in the sea. Conspicuous among these was *Ichthyosaurus*, a fearsome creature with a large mouth as well-equipped with sharp teeth as that of the crocodile today.

During Jurassic times the floor of the sea was subject to occasional undulatory movements. The crests of some of these undulations rose above sea-level and formed long stretches of land on which grew strange-looking trees allied to the Monkey Puzzle *(Araucaria)* and to tree ferns. As yet there were no herbs with coloured flowers, no grasses, no grassy swards. The landscape must have been a drab expanse, varied here and there only by the dark green foliage of the trees. The scene was, however, enlivened by the presence of many kinds of reptiles of all sizes, crawling, running on all fours or leaping like the kangaroo. Some small forms fed on insects. The larger ones fed on foliage or preyed upon their fellows. There were also some with such large bulky bodies that only by wading up to their necks in water could they support their great weight. Other lightly-built reptiles had large flimsy wings like those of the bats. These pursued and fed upon dragonflies and other flying insects which were at home in the air. As yet there were no birds. From all this it is evident that this period was one in which reptiles dominated air, land and sea. There were a few furred animals or mammals, but they were no larger than rats and mice and survived largely because they led a furtive existence in the crevices of rocks or hiding in the foliage of the trees.

The Jurassic Period was brought to a close by a slow, nearly uniform, lowering of the earth's crust in the British area. The sea flowed in and eventually submerged the whole with the exception of the highlands of Scotland and Wales. The waters of this sea were clear and warm and teemed with myriads of minute organisms that made tiny shells no larger than a pin's head. These lived their lives floating in the sunlit waters near the surface. When they died their shells drizzled down to the floor and helped to form deep deposits of white mud which ultimately solidified into the white chalk which makes up the cliffs of Kent and the Downs. The period during which this chalk was formed is known as the Cretaceous (creta = chalk). It was the last of the great divisions of the Mesozoic Era. The thickness of the chalk deposits grew to be 1,000 feet or more and the laying down of the last of those deposits marked the ending of that period and era. The long process of burying the Peak District now ceased.

RE-EMERGENCE

The Mesozoic Era lasted about 120 million years, during which the Peak District lay hidden under an ever-increasing cover of limestones, sandstones, clays and chalk. The time of its eclipse was now, however, drawing to a close, for the British area, after its prolonged period of sinking, began to rise once more. The floor of the chalk sea was uplifted and converted into dry land, an event which marked the opening of the Cainozoic Era.

This uprising did not take place uniformly over the whole area but was most marked along a line running northwards through the area of the Peak District and the Pennines and resulted in the appearance of a low elongated dome. This formed an island which had the scenic features of chalk downs. The crest of this dome served as a natural water divide and many streams flowed down its gentle slopes. The sea floor continued to rise and the boundaries of the land were extended outwards in all directions. The streams lengthened out across the broadening plains as these rose out of the water. Uniting with one another they merged into larger rivers flowing to the sea which by this time was far away.

Wherever streams and rivers flowed they excavated channels which as they deepened were widened into valleys. All this involved the gradual removal of large quantities of rock. The chalk was the first to be attacked. After that the more deeply seated layers of the Jurassic and Triassic rocks were penetrated, a piece of work which was begun and carried farthest upon the original dome. There at last the Carboniferous rocks began to peep out along the beds of the streams. The latter continued to carry away the waste formed by the gradual destruction of the younger rocks on either side of the valleys. But still the streams went on cutting their channels more and more deeply into the Carboniferous rocks. These channels were also widened into valleys and these old rocks came to occupy a correspondingly larger portion of the valley sides. The Mesozoic rocks were steadily reduced to mere remnants capping the hills on either side. Eventually even these disappeared completely and left the Carboniferous rocks in sole possession of the Peak District.

The long eclipse had ended. The precise date for that event is unknown for the complete removal of all the waste from the destruction of so much rock left no tangible evidence behind. It is, however, probable that it happened shortly before or at the opening of the Pliocene period, which was the last of the four major periods into which the last or Cainozoic Era has been divided.

The removal of the Mesozoic rocks was not the only great event that happened during this long period of time. The living garment that clothed the landscape also changed. True flowering plants that produced seeds in closed caskets had come into being in other parts of the world during Cretaceous times. These invaded the new lands as they rose into being over the British area. Among them were many kinds of woodland trees and numerous flowering herbs and grasses which covered the ground as with a multi-coloured carpet. Over this beetles crawled, butterflies flitted and bees hummed. Winged reptiles had passed into oblivion and their place was taken by birds. The long reign of reptiles was over and only a few insignificant representatives remained. On the other hand, mammals were rising rapidly to a position of dominance. A wonderful variety of insect-eaters, vegetable and flesh feeders haunted the woodlands and sported on the grassy plains. Many of these began to show recognisable resemblances to modern and familiar types, to horses, deer, wolves, cats and monkeys. Usually they showed some feature that would seem to us to be peculiar.

The Peak District now entered upon the last phase of its long and eventful history. The carving of the scenery as seen now had begun. Even during the time of its eclipse the blue-print for the general arrangement of its hills, valleys and moorlands was being drawn. When in early Cainozoic times the streams and rivers first began to score the surface of the chalky slopes, their direction of the flow was the same as that in which the Mesozoic rocks were dipping. Just as a saw when it has begun cutting through a log of wood must keep to the same line in which it started, so likewise a stream, having once made a track, must keep to it. Having made its bed it must lie on it.

Thus it came to pass that the arrangement of streams laid down upon the surface of the chalk was eventually incised into the underlying Carboniferous rocks. These, however, had been folded long ago and the layers had been tilted at varying angles in differing directions. Henceforth the streams flowed on regardless of the directions in which the rocks dipped or of the way in which they were folded. Such a drainage system is said to be *superimposed*. One outcome of this is that the scenery along the valleys is more varied than alongside normal streams. Thus, for example, in the main valley of the area, that of the Derwent, grit scars, limestone gorges and wide fertile dales alternate with one another.

THE ICE AGE

One last event remains to be mentioned. The period popularly known as the Great Ice Age was on the whole a time of mild and even warm climate, interrupted now and then by polar conditions when icefields formed over the mountainous regions and, flowing thence, covered the lowlands and filled up the adjoining sea basins. During one of the earliest of these interruptions, or glaciations, all but the extreme south of England was covered with ice. Nevertheless the presence on the South Pennine moorlands of rocking stones and other stones fantastically shaped by prolonged weathering in pre-glacial times indicates that some of the highest parts of the Peak District were never subjected to the scouring action of overflowing ice.

During a later glaciation ice from the Irish Sea invaded the lowlands of Lancashire and Cheshire and, impinging against the flanks of the West Moors, attained an altitude of 1,250 feet. At that time a trickle of ice finding its way across a gap at the Dove Holes, north-west of Buxton, entered the tributary valleys of the Derwent and even reached the vicinity of Matlock. This ice did not, however, leave any appreciable impression upon the scenery of the district. Nevertheless, whenever such arctic conditions prevailed they intensified the action of more normal agents. The snow which accumulated in the winter melted in the early summer. On the shale areas the ground became sodden in the daytime and frozen at night with the result that soil creeping and land sliding took place on a massive scale and greatly accentuated the concavities of the valley slopes.

Elsewhere in England and north-west Europe, Old Stone Age Man appeared upon the scene and gained a livelihood by hunting. Outside the Peak District, though quite near, the caves at Creswell have yielded a very full and unique record of his sojourn in this part of England. In striking contrast is the solitary discovery of one flint implement referable to that time within The Peak in the vicinity of Wirksworth. It is sufficient, however, to indicate that Palaeolithic Man was an occasional visitor. As will be seen later, evidence of prehistoric man is more abundant in post-glacial times.

REFERENCES

Marshall, C. E. *Guide to the Geology of the East Midlands*. University of Nottingham (1948)

Memoirs of the Geological Survey. H.M.S.O.

Wray, D. A., Edwards, W., and Trotter, F. M. *The Pennines and Adjacent Areas*. Third edition. British Regional Geology. H.M.S.O. (1954)

Waters, R. S. and Johnson, R. H. The Terraces of the Derbyshire Derwent. *East Midland Geographer: 9* (June, 1958)

Downie, C. and Neves, R. (editors). *Geological Excursions in the Sheffield Region and Peak District National Park* (1967)

3. The Making of the Scenery

Approach we then this classic ground:
More gentle name was never found
By chance, nor more of picturing sound
To tell the spirit of the scene.

Edmund Blunden: Dovedale

The scenery of the Peak District is a treasury of features, some new, some old. Relics of the far-distant past are closely linked with others of more recent origin; some beautiful, others rugged, weird, mystifying. It is the outcome of a long process of sculpturing which was produced by slow incessant change. Throughout all these ages, however, the rocks have remained unchanged in quality and arrangement and have exerted a constant though passive influence upon the landscape. A brief consideration of the nature of that influence will provide a useful background against which to watch the development of the scenery as it exists today.

There are three main types of rocks in the district – limestone, grit and shale. Volcanic tuffs and intrusive dolerite play a minor but interesting part.

ROCKS AND SCENERY

Limestone is a hard, almost impervious stone. Consequently very little rain-water penetrates into its substance. This fact shields it from the shattering action of frost. On the other hand, it is slightly soluble in natural water. Just as a cube of sugar becomes rounded as it dissolves in tea, so the contours of the limestone surface tend to develop smooth curving outlines.

Even the purest limestone contains small quantities of dust and other earthy materials. This is left behind on the surface of the rock and accumulates to form soil. On slopes the soil is usually too shallow to give firm foothold for trees, except for the ash, but it is covered by a carpet of grass and flowers which provides valuable pasturage. The more level ground on the platforms and along valley bottoms has a covering of much deeper soil. Here trees and bushes may flourish except where the altitude is such as to expose them to the influences of strong winds.

The debris shed by the vegetation becomes mixed with the soil. There it rots and produces weak humic acids which are taken into the water percolating down through the soil and thus intensify its solvent action upon the limestone.

Limestone, like other rocks, is arranged in layers or strata, each of which is broken into more or less cubical blocks by the presence of two sets of cracks known as joints and bedding planes. Descending water finds its way down even the finest of the joints and along the closest bedding planes, and by dissolving the stone on either side widens these into open fissures. Eventually these become so spacious that most of the rain-water abandons the streams upon the surface and flows away along these newly formed underground channels. Below a certain depth all cracks and fissures are permanently filled with water. That depth is spoken of as the water-table and thus becomes the surface along which these underground streams flow.

Sometimes a surface valley is deep enough for its floor to lie along or even slightly below the water-table. In these circumstances a surface stream or river is maintained which is remarkably constant in its flow and rarely swells into flood in wet seasons or shrinks and disappears in times of drought. The Dove is a good example of this type of river.

Grit, on the other hand, is practically insoluble in water. It is, however, so porous that rain-water is quickly absorbed and, soaking inwards away from the surface, leaves this dry. The water, however, fills all the pores and cracks in the deeper layers, which then become saturated. The dry superficial parts of the grit are consequently less liable to destruction by frost action and therefore retain their angular forms and sharp edges. Some grits contain grains of felspar, a mineral which is gradually rotted by natural water. In these cases the rock disintegrates more rapidly.

Shale is more finely porous than grit. Water therefore percolates into it more slowly and is retained in its surface films. There its prolonged presence softens the rock and favours the pulverising action of frost. In areas where shale is the dominant rock, vertical erosion by streams and valley-widening proceed more rapidly. In these areas landslips occur on the steeper slopes and under the overwhelming pressure of superincumbent rocks the shaley sides of the valleys may even begin to bulge. Examples of such features may be seen around Edale and other vales.

The limestone, which is the oldest rock in this district, underlies all the others and occurs as a great mass at least 1,500 feet thick. In its uppermost portions thin layers of shale appear. Passing upwards these thicken and the intervening bands of limestone become thinner and eventually disappear. The succeeding 1,000 feet of shales are called the Edale Shales.

Thin beds of grit begin to appear in the upper levels of the shales which are referred to as the Grit Shales. Passing upwards through the rock series the grit layers become more massive. Of these there are five, of which the Kinderscout Grit may be specifically mentioned, for in the north of the district around Black Hill and Bleaklow it attains a thickness of as much as 600 feet. All these grits, however, have a lenticular form and tend to become thinner towards the south where the surface features they form are much less prominent than in the north.

Turning now to the arrangement of the rocks it will be recalled that they have all been folded, a fact which has a marked influence upon the features of the landscape. The folds are of the two general types – upfolds or anticlines and downfolds or synclines. The sides of each fold are its limbs and these link the crest of the former with the trough of the latter. The rock layers are horizontal or nearly so at these two points but are more or less steeply dipping in the limbs.

Erosive or denuding agencies naturally attack the upfolds to begin with and in doing so strip away the younger rocks first and the older ones in turn until in the centre of the crest the limestone is brought to view as in the Derbyshire upland. Northwards from the upland, earlier stages in this stripping process are exemplified in succession. In the moorland area much of the grit cover still remains. On the other hand, on the downfolds the youngest rocks survive longest in the centre of the trough, hence the presence of Coal Measures in the lower Goyt valley. Along the zones of country occupied by the outcropping limbs of the folds the rocks dip down from the surface and the edges of the strata produce such scarp-like features called 'edges', as Axe Edge, Froggatt Edge, Baslow Edge and Black Edge (Fig. 4).

TIME AND SCENERY

The Peak District is but a minute portion of the earth's crust. This latter is often spoken of as 'terra firma' as though it were quite rigid. But it is rigid only in the sense that a block of wood is rigid. Anyone who has jumped off a diving-board knows, however, that when the block is so long that it becomes a plank it is springy and flexible. The rock layers which make up the Peak District are similarly flexible. Deep down in the earth beneath them lies a plastic foundation which during long periods of time has crept slowly from one region to another and the crust of rocks which rests upon it has risen and fallen accordingly. In the middle and again in late Pliocene times movements of this kind took place in the Peak District and resulted in a general uplift of the region. Each uplift probably took place in a series of stages but, for the sake of brevity with clarity, only the total results will be considered for the two occasions.

With each total uplift of the Peak District its level above the sea was increased. Consequently all the agencies which had almost gone to sleep were aroused into activity once more; water flowed more rapidly; vertical and later on horizontal erosion were renewed. Thus it came about during the remaining 15 million years of Pliocene, Pleistocene and recent times, that rain and rivers, frost and glaciers gave to the district those magical touches of beauty which make it so attractive to its many visitors.

The process of carving the surface was by no means haphazard. On the contrary, as the result of successive uplifts there was a majestic rhythm about its progress which inscribed its score everywhere. For those who learn to read that score the appreciation and enjoyment of the scenery are greatly enhanced. How was that score written? That is the next problem to be explored.

The key agents in carving inland scenery have always been the rivers and streams. Those of the Peak District are all tributaries of large rivers – the Mersey, Trent and Don – which in turn pour their waters into the sea. This last phrase, though trite, leads straight to the central influence that ultimately controls the whole activity of running water, for both flow and erosive activity cease at sea-level. This last term, however, lacks precision for both land and sea rise and sink independently of one another. Strictly speaking, therefore, the controlling influence is the relative level of these two and this is called the *base level.* Throughout its course a river flows and works only so long as its bed is above base level.

The character of the work done by the river changes at different portions of its course. In its upper reaches the bed is steep and water flows rapidly. Like a man running swiftly, it follows almost a straight course and overcomes all obstacles that lie across its path. Downstream, as the slope of the bed becomes more gentle the rate of flow declines and the water is more easily diverted. Henceforth the river meanders from side to side.

Upstream the more rapid flow gives to the water greater power for rolling boulders and stones along, and these by their continual passage wear grooves and channels across even the hardest rocks. This wearing of the river bed is described as *vertical erosion*. Downstream this power is lost. Nevertheless as the river meanders along it impinges against the outside bank of each bend. In times of flood, when the water is carrying a load of sand and gravel, it undercuts and wears the bank away. It is then said to be *eroding horizontally*.

Upstream again, the valley sides come down to the margin of the water and the valley is V-shaped in cross-section. Downstream, on the other hand, as the river meanders it also erodes the lower fringes of the valley sides and thus the valley bottom is widened into a flat which increases in width as it approaches the sea. This broad flat with its cloak of gravel, silt and mud constitutes the alluvial plain.

While all these changes are going on frost and rain are busy working upon the valley side. The one pulverises the rocks and converts them into soil. The other washes away the surface of the soil and discharges it into the stream. At the same time the rain soaks into the soil and causes this to swell. In dry weather the water evaporates and the soil shrinks. This alternate wetting and drying, swelling and shrinking causes the soil to wriggle slowly downhill and ultimately to fall into the stream. This process is known as *soil creep*. Long after the river ceases to erode vertically these processes continue, the valley sides recede from the stream and the valley increases in width. There is, however, a limit to all these activities; they cannot be carried on below the level of the alluvial plain which serves as their local base level. As the steeper convex sides of the valley recede from the river they leave behind them a gently sloping apron of deep fertile soil which in these days is occupied by prosperous farms and villages.

Meanwhile the plateau-like high ground between the valleys diminishes in height and extent and becomes reduced to a gentle rise of ground. In this way hilly and even mountainous country is levelled down almost to a plain, a 'peneplain' (pene = almost), and is characterised by a gentle undulating surface with broad

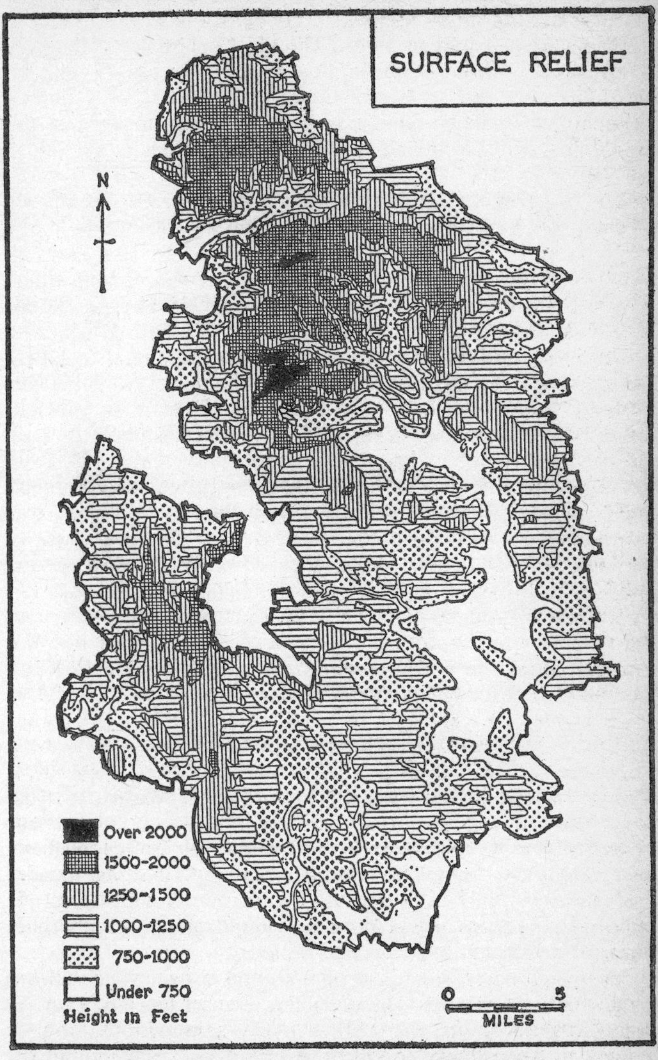

Fig. 5 Surface relief of the Peak National Park

open valleys and low spreading rises. This late stage in the development of landscape was attained in the Peak District in or about early Pliocene times. The level of the land relatively to the sea was about 1,200 feet lower than it is now. It remained near this level for so long a time that the work of denuding agencies was carried almost to complete fruition even in the uppermost and far inland portions of the drainage system in the region.

The open breezy highlands of the Peak District are remnants of the peneplain then produced. The features described above are best exemplified in the limestone uplands where the rock is almost uniform in quality. An excellent viewpoint from which to see them in profile is from the summit of Thorpe Cloud. In the moorland areas in the north the alternating grits and shales have produced a rugged surface. Upstanding peaks were absent, but even in those far-off days the plateaux and ridges rose above the general level of that ancient peneplain. Ancient indeed, for it dates back to early Pliocene times, when for several million years land and sea remained relatively stable with only small oscillations of level. The denuding processes continued their work without serious interruption until it was accomplished. At that time the Peak District did not rise above the surrounding country as it does now, for its surface was only a small part of an extensive peneplain that sloped away gently towards the far distant sea, the sea whose nearly constant level had for so long a time exerted a controlling influence that was felt along the whole length of every river and stream and across the breadth of the whole countryside.

BEYOND THE DALES

For many tourists the word Derbyshire spells 'dales'. It is the dales they love to explore. It is up the dales they hike. When eventually they emerge into the upland, as for instance out of Lathkill Dale they lose interest for 'there is nothing to see', nothing but stone walls and tame pastures. Nevertheless, for a full appreciation of the more exciting vistas within the dales the story of these uplands must be told.

The sequence of events detailed in the last section is known as a *cycle of erosion.* Whenever the level of the land rises or that of the sea sinks, the current 'cycle' is interrupted or even ended. A general movement of this kind took place about the middle of the Pliocene period. The whole landscape was uplifted about 300 feet and remained at the new level for a long time. The rivers were rejuvenated and recommenced excavating their channels, first of all in their lower reaches. The point near which the steep new bed joins up with the gently sloping old one is commonly called the *knick point*. Below this the newly formed valley was at first a narrow gorge and ran like a trench along the floor of the broad, open, ancient valley of the former cycle. Above the knick point that valley remained unchanged. The river continued excavating its channel and the knick point receded upstream along nearly the whole length of the former valley.

Meanwhile the sides of the gorge were worn by weathering agencies into steep and then gentle slopes. The gorge was thus slowly converted into a wide valley lying within the limits of the old one. Along its margin where the steeper side of the new valley merged into the floor of the old one there was a 'break of slope', essentially a greatly elongated extension of the knick point. Had this widening process been carried on to its utmost limit all traces of the older landscape would have been destroyed. Fortunately this cycle of erosion was interrupted in late Pliocene times and consequently relics of the earlier landscape survive in the loftiest portions of this upland.

The late Pliocene uplift raised the general level another 200 feet and all the weathering machinery was set going once more. Throughout the Pleistocene and later times, river channels were worn deeply once more and thus the dales as now seen came into being as the youngest features in the Derbyshire scenery.

That, then, in rapidly drawn outline is the general story of this limestone scenery in the Peak District. Further details must now be considered and these vary from dale to dale mainly in association with the size of the streams.

The Lathkill, though only a small stream, provides a pocket edition of the whole story. In the centre of its basin lies Monyash surrounded by a broad open valley formed mainly in mid-Pliocene times. The high ground enclosing this basin bears the last traces of the early Pliocene peneplanation. Downstream from Monyash the floor of the valley is gashed by the dale, the excavation of which was begun in late Pliocene times and continued until now.

The survival of so many traces of the early phases in the development of the landscape is largely due to the fact that it lies wholly within the limestone region. Apart from the Lathkill there are no surface streams. Had such streams been present they would have inscribed an intricate pattern of new valleys and in doing so would have removed still further and larger portions of the ancient surface. The rain, however, instead of flowing off along the surface descended down cracks and joints in the rock and dissolved out underground channels along which it journeyed to the newly forming dale. Thus many of the dales are now dry at the surface, Gratton Dale being a good example.

It must not be supposed that the relics of the more ancient landscape have survived the passage of ages without undergoing change. On the contrary the whole of the limestone area is like a marble statue which has been exposed to the weather for a long time. Every fall of rain which has washed its surface has dissolved away some of the marble and gradually destroyed the finer details of the carving. But the major features of the face, chin, nose and eyes can still be recognised as such. So though the limestone of this upland has been washed by rain for 5, 10 or 15 million years, the major features originally carved upon its surface can still be recognised.

The water which disappears underground is by no means idle. It finishes its downward journey when it reaches the water-table, that is to say the surface below which all cracks and joints are already filled with water. This surface differs from that of a lake in that it is not flat and horizontal but has a general slope roughly parallel to that of the surface of the countryside above it. The water newly arrived from above now flows along this watery surface, dissolving the limestone away on either side of the crack which forms its path. The level of the water-table rises and falls according as the season is wet or dry and consequently the crack is widened into a tunnel or cave within the limits of that rise and fall. Such a cave may be seen soon after entering the dale from its upper end. In a rainy season the water-table rises above the floor of the dale and the stream is then seen issuing from the cave as a surface stream. In times of drought the water-table sinks and the stream, finding that its underground channel is enough, disappears from the surface and continues its course below ground. With this alternating rise and fall of the water the roof of the underground channel is being gradually dissolved away and ultimately collapses and thus a new and romantic addition is introduced into the floor of the dale. This type of deepening has been going on in all the dales since late Pliocene times.

The limestone varies in quality from place to place throughout the uplands. Sometimes it is rich in fossils which are slightly less soluble than the rock itself. This is particularly the case where coral or shelly reefs existed on the floor of the Carboniferous sea. At such places the rock is less rapidly dissolved than is the surrounding limestone and so it stands up as a more or less prominent hill known as a reef knoll. Thorpe Cloud, Bunster, Wetton and Gratton Hills are examples of such knolls.

One important result accruing from the solvent action of rain-water upon the limestone is that the insoluble residue remains on the surface and accumulates to form soil. This is particularly the case where the surface is almost level, for then rain-wash and soil creep have only slight effect. The soil is then said to be stationary. Where the surface has a marked slope these two factors come into action more vigorously and the soil is transported downhill. On such slopes the soil covering is thin but in the adjoining valley it is deep. These differences from place to place exert an important influence upon the agriculture of the upland.

The wet surfaces over which the escaping water flows become moss-grown. The moss, however, takes up the carbon dioxide from this water with the result that it can no longer hold much lime in solution. The latter is therefore deposited and, covering the moss, forms a spongy-looking rock called tufa. This is often used for rockeries and sometimes for building as in the case of Tufa Cottage in the Via Gellia. In some springs the water rises from great depth and as it does so the pressure upon it diminishes rapidly. Once more much of the lime held in solution is set free and covers any objects, such as toys and birds' nests that are put into it, with a hard coating of lime. As they have the appearance of having been changed to stone such springs are spoken of as petrifying wells.

In caves the water dropping from the roof or flowing down the sides also parts with its lime. In doing so it deposits this either as icicle-shaped pendants from the roof or as pinnacles rising from the floor. These features are known as stalactites and stalagmites respectively. Sheets of lime may also be laid over the walls or hang like curtains from ridges on the ceiling. When such a cave is judiciously illuminated it becomes a beautiful scene and a profitable centre of attraction.

WITHIN THE DALES

The rambler approaching Dovedale and the uplands along the Belper-Ashbourne road begins to catch visions of the promised land when he passes beyond Hulland. He looks across a succession of level-topped hills with an altitude of six or seven hundred feet. These level tops are relics of a mid-Pliocene peneplained platform. Away in the distance beyond them the uplands appear as a lofty rampart, bounding this platform on the north, and extending to the Weaver Hills.

In the centre of the rampart is Thorpe Cloud, rising like a bastion at the entry to Dovedale. Making that his first objective, he scrambles to the top and looking northwards, finds himself on a level with the upper or early Pliocene platform and sees it as a gently undulating landscape having all the general characteristics of a peneplain.

In striking contrast to that is the deep steep-sided dale looking like a cleft in that ancient landscape. Closer inspection of the cleft reveals a feature that is easily overlooked. The precipitous sides of the cleft do not rise to the level of the upland itself for their rims spread out like a shallow funnel about 200 feet deep. The sides of this funnel curve upwards from the lips of the gorge to the level of the upland platform. The funnel is in fact the profile or cross-section of a moderately broad valley in the floor of which the gorge has been carved. This floor extends southwards into the 600–700-foot platform already noticed between Hulland and Ashbourne. The sight of this valley takes the thoughts back to middle Pliocene times when the Dove flowed along this floor and debouched on to the plain of which this platform is a survival.

Leaving his eyrie, the rambler descends and sets out to explore the dale. At first his path lies alongside the stream. Presently it rises steeply and takes him up to the Lover's Leap, the name given to a spur which projects towards the gorge. Standing upon the tip of the spur he looks down into the gorge and his eyes come to rest on the wooded slopes of the opposite side. Here and there amongst the greenery may be seen grey pinnacles of limestone known as the Twelve Apostles. How these came to be there will emerge later. For the present it must suffice to say that they are the degraded remnants of another spur that once projected from the other side, the counterpart of the Lover's Leap.

Turning back from viewing the Twelve Apostles, it is seen that the flat surface of the Leap slopes upwards like a valley side and merges into the upper plain. Looking up the dale similar spurs may be seen farther on. Each of these shows the same traces of the old valley features, for all these spurs are also relics of the mid-Pliocene valley. The observer is standing where at that far distant date the Dove actually flowed.

Leaving the Lover's Leap behind, the explorer descends past the successive levels through which the river excavated on its way down to its present bed. Once more the path lies alongside the stream through smooth grassy flats unimpeded by boulders. On either side are the rocky cliffs criss-crossed by vertical cracks or joints and horizontal or gently dipping bedding planes. In winter some of these become filled with water from melting snow. When this freezes it expands and the ice, acting like a quarryman's wedge, gradually prises lumps of rock from the face of the cliff which fall and form a blocky scree at the base.

Here and there other narrow spurs once projected into the dale, but these have been partly or wholly destroyed. Frost working at both sides of the spur has worked its way in along major joints or fractures and cut it up into isolated columns such as that of the Ilam Rock. Rain falling upon such a column dissolves the corners and edges and eventually reduces it to the shape of a pinnacle such as those already seen from the Lover's Leap.

Another feature of the dale that is easily overlooked is the fact that it follows a winding course. One result of this is that the vistas are usually not long but are closed in by a succession of rocky pictures often of great beauty. As each bend of the gorge is passed there come into view new cliffs, fresh and fantastic shapes or even a cave, such as Dove Hole and Reynard's Cave. These last are a reminder that the excavating of the dale has not been entirely due to the direct deepening of the river channel. Disappearing streams like those of the Lathkill and the Manifold have played their part by dissolving underground passage-ways beneath the floor of the dale which became open to the air and the sunshine, in the way already described for Lathkill Dale, and henceforth became part of the main dale. The two caves just mentioned were formed by tributary streams which, however, were drained dry when the Dove itself deepened its channel below their level.

The winding of the dale is reminiscent of the meanders of a river flowing along an alluvial plain. When such a river is rejuvenated the knick point, followed by the deepened channel, travels upstream along the meandering course round bend after bend. Such was the history of the Dove when, with the mid-Pliocene uplift, it became rejuvenated and incised its new valley into the early Pliocene peneplain, and at a later date excavated the dale in the floor of that valley.

At last the explorer comes to the end of the dale and finds that it opens out into a normal valley. From this point up to its source on Axe Edge it does not flow over limestone but over shale with some grits. As already seen these shales yield much more readily than does limestone to the destructive influence of rain and frost. When therefore this uppermost section of the river came under the influence of the latest rejuvenation, the sides of the deepening channel were more rapidly worn back and a wide valley formed.

THE MOORLANDS

The striking change of scenery met with along the upper reaches of the Dove is a reminder that the limestone uplands form only part of the Peak District; the other part is made up of grits and shales. The change reflects the still greater contrast between the uplands and the moors. This gives to the district much of its attractiveness as is typified by the streams of visitors to the dales and the comparative trickle of energetic hikers who make the lofty, windswept moors their main objectives.

The rocks which form the moorlands completely encircle the uplands like the oval frame of a beautiful portrait. At the top of the frame, where the grits have their maximum thickness, the moulding stands out in bold relief. Along the sides this declines and at the bottom it almost disappears for the grits become much thinner as they pass southwards round the limestone. On the south such features as they produce still remain largely hidden under the ancient covering of Triassic sands and marls.

In the uplands the limestone is almost uniform in character, but in the moorlands extremes meet and resistant grits lie in close juxtaposition with the much more easily weathered shales. This is illustrated in The Peak itself which is capped with a mighty slab of grit, several hundred feet thick, resting on a pedestal of shale. The slab forms a plateau bounded by steep rugged cliffs. The pedestal is flanked by slopes whose surface is roughened by landslips or broken by steps caused by the presence of occasional thin beds of grit.

The rock of the plateau is so porous that rain-water soaks into it rapidly and, percolating downwards, saturates the lower layers. This great slab of grit therefore functions as an underground reservoir having an impervious floor but is without retaining walls around its margins where, at the junction of the two kinds of rock, the water leaks slowly away and is lost. Within the grit the water flows freely through the fissures but elsewhere its flow is hindered as it seeps slowly through the fine pores of the rock. Unlike the water-table of an ordinary open-air reservoir, the surface of which is perfectly flat, that in the grit is heaped up in the centre and slopes thence in all directions.

Wherever a valley has been developed on the plateau deep enough for its floor to reach the water-table, water escapes and forms a stream flowing along the valley bottom. Elsewhere the plateau is dry and streamless.

At the base of the cliffs bounding the plateau water leaks away perpetually and, soaking into the adjoining shale, softens it to such an extent that it yields to the pressure of the overlying rock and is slowly squeezed out. The cliff, with its foundation thus weakened, eventually collapses and tumbles its fragments down the slopes. It is in this way that the slabs of rock which cap the highest grounds in the district have been and are being worn away gradually, so that large plateaux in time become small ones, e.g. Brown Hill and Stanton Moor, and small ones become conical peaks or pikes such as Win Hill near Hope and Oaker Hill near Matlock.

Where a crack or fissure running through the rock emerges at the cliff-face a copious spring of water gushes forth. At such a point the destructive action just described takes place more rapidly and leads to the formation of a notch, a gully or even a 'clough', the floor of which is cumbered with fallen blocks of grit. All these features may be seen well developed round the margins of any grit plateau or along a grit edge.

Turn now to the pedestal which supports the rock. This consists mainly of shale with occasional thin grits. Each of these as it crops out to the surface acts as a protective covering for the shale below; but the shale above it is weathered into a concave slope. This merges downwards on to the flat upper surface of the grit which is exposed as a narrow platform.

The vicinity of Matlock Bank yields an excellent and easily accessible example of these types of topography. Standing in Salters Lane and looking across the valley, the step-like features produced by the sub-divisions of the Kinderscout Grit are well seen. They are indeed emphasised by the plan of the town in which main roads run parallel to one another along each shelf and are flanked by buildings which have developed along the ledges.

On the skyline above all this are Matlock and Farley Moors, which owe their presence to a small plateau formed by the Chatsworth or Belper Grit. This layer, however, does not lie quite flat but is bent into a broad shallow downfold having a slight southwards tilt. Rain-water falling on this seeps through the rock to the centre of the fold, where it accumulates as a valuable underground reservoir from which the Matlock Urban District draws its main water supply.

The vegetation cover of the plateau shows an interesting zonation of plants from those which flourish on very dry situations near the scarp margin where the ground level is high above the water-table, to those which favour the boggy conditions near the centre where the ground has dipped down to the level of the water-table.

On the southern margin of the fold the water from the reservoir spills out as a copious stream, the Bentley Brook. This flows for a short distance through a rock-strewn clough whose rugged sides suddenly diverge and pass into two long curving scarps which enclose a broad shallow valley excavated out of the underlying shale down on to the upper surface of another grit layer. This is a good example of a type of hanging valley which is frequently met elsewhere in the moorlands. After flowing for about a mile the stream plunges over the lip of the grit and sets out on a tempestuous journey down the steep valley side east of Matlock, cascading through Lumsdale in a succession of cloughs and ultimately joining the Derwent river.

Similar combinations of topographical features are met with repeatedly throughout the moorlands especially in its northern section. Here, as already seen, the underlying form of the region is based upon a dome-like arrangement of the rocks. Owing to the asymmetry of the anticline its crest lies nearer to the west side, a fact which accounts for the asymmetric position of the dominating features situated along the line from Kinderscout to the extreme northern end of the Peak District, where the influence of the Derbyshire dome ceases to be felt. There the moorlands have a minimum width of only about two miles as contrasted with twelve or more miles in the latitude of The Peak. Along the crest the higher members of the Millstone Grit series have been removed and the lower ones which are more massive dominate the scene. In association with the dip of the rocks away from the crest, the summit level declines most rapidly towards the west. Towards the north-east and east, owing to the low angle of dip, the grits and shales give rise to more widely spaced scarps and broader vales than those on the west.

Thus the general form and arrangement of the major features of the moorlands are controlled by the arrangement of the rocks. But these alone would have led to nothing more than a vast expanse of desolate and even repulsive moors. The area has, however, been redeemed from such a fate by the carving activities of running water. Everywhere are to be found streams fed from the inexhaustible reservoirs of the grits. These set out from the crest on their journey to the lowlands, carving for themselves narrow, often deep gorges through the massive grits. By joining forces they become larger streams. With their power increased and aided by associated agencies they excavate an endless variety of cloughs ranging from mere notches down the faces of steep scarps to narrow, wild and impressive valleys such as that of the Crowden Great Brook which opens on to the north side of Longendale. A swift torrent flowing down the western slope of Kinderscout starts as the Kinder Downfall, which is the only considerable waterfall in The Peak. Two rivers, the Etherow and the Derwent, are the centres towards which many of these streams converge. The spacious valleys of these rivers have steep sides with the usual step-like grit features at successive levels crowned by magnificent scarps. In each case the left side rises more abruptly, forming a continuous feature of great grandeur only slightly broken by notch-like cloughs. On the right side the valley sides are less abrupt and are deeply dissected by larger valley-like cloughs.

In the northern outskirts of the drainage basin of the Derwent much of the Kinderscout Grit has been removed and has left the underlying shales with minor grits exposed over extensive areas. These are occupied by moors and mosses lying some 500 feet lower than The Peak. They also are redeemed from monotony by the presence of attractive tributary valleys, spacious but steep-sided.

East of the Derwent the outcrops of the Kinderscout Grit continue south as the East Moors. These are defined on the west by a series of prominent scarps (Derwent, Stanage, Froggatt and other Edges) overlooking the main valley. They dip at a steeper angle than that already seen in the north but carry a similar series of dip slopes and scarps. Here, however, the continuity of the latter is interrupted by the transverse folds already mentioned. They give rise to structural and surface features like those described above in detail for Matlock Moor.

In the West Moors, between Buxton and Macclesfield, the geological structure of the district reaches its maximum of complexity in sharp north-to-south folds crossed by minor ones trending west to east, thus producing an area of intermingling types of scenery like those already described but on a smaller scale, surrounding or enclosing in their synclinal hollows patches of lowland types due to the presence of pockets of Coal Measures and even of Triassic rocks.

One more element in the make-up of the district remains to be mentioned. The Edale Shales which lie between the limestone and the grit have a maximum thickness of some 1,000 feet. Under normal circumstances it should crop out as a relatively broad zone between the limestone uplands and the moorlands. Owing to cross-folding and faulting, especially in the west, the zone is broken into a number of separate patches which give origin to the gracious landscapes of Darley Dale, Edale, overlooked by Mam Tor, and Hope Dale. The more extensive patches lie outside the area to the south and southwest.

THE WORLD UNDERGROUND

Reference has already been made to the formation of caves as a common feature of the limestone area of The Peak. The exploration of these caves under proper auspices is a challenging form of recreation for the physically fit, exciting but rigorous, while many of them are of special scientific and archaeological interest. Some of the larger and more spectacular examples are exploited commercially, like the famous Blue John Cavern (strictly the Blue John Mine), the Peak and Speedwell Caves and the Treak Cliff Caverns, all in the Castleton district, and are an unfailing attraction for the general sightseer, though they are not quite so impressive as those of Cheddar.

While the processes by which caves are formed are not disputed, opinions differ as to the conditions necessary for the processes themselves to operate. As already explained, caves mainly owe their origin to water action, either as a solvent in the limestone or as an agent for the transporting of rock material by which the beds of underground streams are scoured. The relative importance of each of these forms of water action doubtless varies with local conditions. Authorities hold different views, however, as to where, in regard to the underground water, the cave-forming processes take place. Some maintain that caves originate and develop above the water-table within what is termed the 'vadose' zone, i.e. the zone between the surface and the level of saturation, within which water moves downwards by gravity. This view implies that water percolating downwards from the surface will dissolve almost all the calcium carbonate it can hold before it reaches the water-table. Others have shown that many caves must have originated below the water-table in the 'phreatic' or water-logged zone. Regarding this issue much depends on such factors as the extent to which solutional activity may continue below the water-table and the nature of water movement resulting from pressure exerted in the phreatic zone. It is likely that each of the theories is applicable to certain caves. In many cases, since there is ample evidence in the Peak District, as elsewhere, of past changes in the level of the water-table, it is reasonable to favour a compromise involving the application of both theories.

Although the phenomenon of limestone caves has been widely studied in many parts of the world, there is scope for much further investigation. For this the Peak District presents an obvious field. As Dr G. T. Warwick has pointed out, the need is for a careful examination of particular caves, including the basic work of surveying them, without which detailed study cannot be advanced. In this connection the recent work by Dr Trevor Ford on the Treak Cliff Caverns is to be welcomed not only for its intrinsic value but as pointing the way for further investigations.

Caves are seldom found in the Millstone Grit and when they do occur they generally take the form of narrow fissures resulting from the displacement of large blocks of rock. The Kinderlow Cavern on the western edge of Kinderscout is of this type, where a large but narrow block of gritstone has slipped from the main mass, yet still leans upon it, so forming a roof. The Kittycross Cave in Bradwell Dale, though primarily a limestone cave, is partly developed in decomposed toadstone.

In the limestone area it is important to distinguish between natural caves on the one hand and the underground passages and chambers resulting from lead-mining operations on the other, although the latter have often been the means of revealing some of the deepest and most impressive of the natural caves. A good example is the Bottomless Pit in the Speedwell Mine at Castleton which is due to the solution of lime-bearing minerals, mostly calcite, surrounding the ore body. Again, in the Blue John Mine and the Treak Cliff Mine the search for

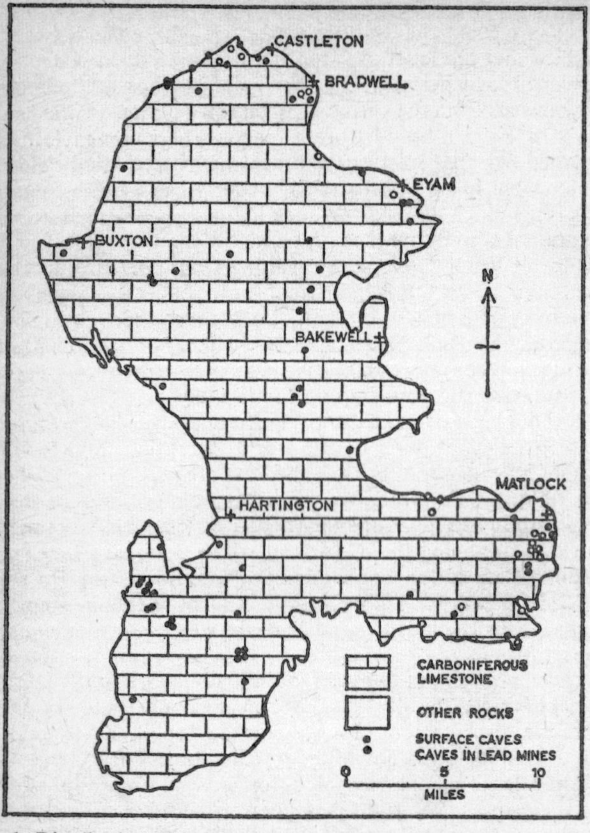

Fig. 6 Distribution of the principal caves in the Peak District. (Based on G. T. Warwick)

lead resulted in the discovery of extensive natural chambers. In fact many of the more intricate caverns now exploited as showplaces owe their accessibility to former mine workings.

In the Peak District the distribution of caves is by no means haphazard. The majority tend to occur around the margin of the limestone in the neighbourhood of streams which flow on to that rock from the higher gritstone areas around it (Fig. 6). Such streams, on encountering the limestone, have developed considerable underground drainage, the more so since much of the limestone, especially along the northern and western margins, is of the reef type, which is relatively pure and highly soluble. Under such conditions there is a marked concentration of caves in particular districts. Of these the Castleton, Bradwell, and Eyam-Middleton districts in the north, the Dove-Manifold area in the south-west and the Matlock-Wirksworth district in the south-east are the chief. In the interior of the limestone area caves occur along some of the valleys such as the Wye and the Lathkill.

Most of the caves, except some in the Castleton group, are situated on the valley slopes or near the present stream level. High-level caves like the Harborough Cave near Brassington, at over 1,000 feet, are seldom found though they are of interest in indicating that ages have elapsed since the water-table stood at such an altitude and they must therefore be of great antiquity. Eldon Hole and Nettle Pot on opposite flanks of Eldon Hill are the only Derbyshire pot-holes, i.e. caves with a vertical pitch, like Gaping Gill in Yorkshire. Eldon Hole is 120 feet long and about 20 feet wide; it reaches to a depth of over 180 feet where it opens into two distinct caverns. Many of the Peak District caves have yielded significant palaeontological and archaeological remains. Thus, from a fissure in the old Victory Quarry north of Buxton, remains of Pliocene mammalian species, including the sabre-tooth tiger, mastodon and southern elephant were brought to light at the beginning of the present century and rank among the few instances of Pliocene cave-finds in Europe. Numerous relics of primitive Man dating as far back as the Bronze Age and even earlier have been found in such places as the Harborough Cave, Thor's Cave, Thor's Fissure and Beeston Tor (St Bertram's) Cave. None of these, however, is of such outstanding importance as regards evidence of prehistoric conditions as the famous series of caves in the Magnesian Limestone at Creswell in east Derbyshire well beyond the boundary of the National Park. It is their detailed features, including calcite curtains of varying hue, cavities lined with fluorspar, diverse forms of stalactites and stalagmites and the subterranean streams which give to the caves their popular appeal. The Peak Cavern has been famous for centuries. It is referred to as a 'marvel of England' in Henry of Huntingdon's *Historia Anglorum,* which was written in the twelfth century. Just as generations ago the Cavern earned a reputation among travellers as one of the wonders of The Peak, so today thousands of people each year are attracted to this and other caves which are exploited as commercial ventures. The caves in this category at present open to the public are the Blue John Mine, the Peak Cavern, the Speedwell Cavern and the Treak Cliff Caverns, all in the neighbourhood of Castleton, the Bagshawe Cavern near Bradwell, the Masson and Rutland Caverns at Matlock Bath, and the Fern Cave and Roman Cave at High Tor, Matlock.

REFERENCES

Trueman, A. E. *The Scenery of England and Wales.* Gollancz (1938)

Stamp, L. D. *Britain's Structure and Scenery.* Collins: New Naturalist (1946)

Miller, T. G. *Geology and Scenery in Britain.* Batsford (1953)

Cullingford, C. H. D. *British Caving.* Routledge (1953). For details of the Derbyshire caves see contributions by G. T. Warwick and others.

Clayton, K. M. The Denudation Chronology of the Middle Trent Basin. *Trans. Inst. Brit. Geographers* (1953)

Linton, D. L. Chapter 2, Geomorphology, from *Sheffield and its Region.* British Association for the Advancement of Science (1956)

Kent, P. E. Triassic Relics and the 1,000-foot Surface in the Southern Pennines. *East Midland Geographer: No. 8* (1957)

Ford, T. D. *The Story of Treak Cliff Cavern* (1957)

Johnson, R. H. Observations on the Stream patterns of some Peat Moorlands in the Southern Pennines. *Mem. ad Proc. Manchester Lit. and Phil. Soc.: vol. 99* (1957-58) and Some Glacial, Periglacial and Karstic Landforms in the Sparrow-Dove Holes Area of North-Derbyshire. *East Midland Geographer: 28* (1967)

Dearden, J. Derbyshire Limestone: its removal by Man and by Nature. *East Mid. Geog.: No. 20* (1963)

4. Climate and Soils

'Tis the hard grey weather
Breeds hard English men.
C. Kingsley: Ode to the North-East Wind

Climate and soil more than anything else determine the appearance of the countryside. Away from the towns the landscape is clothed with vegetation of various kinds, either natural or cultivated. Nowhere is this more apparent than in the Peak District, which exhibits types of vegetation ranging from woodland, heath, moorland and bog to agricultural crops and grass. To understand the nature of the plant cover of our area it is necessary, therefore, to present at least a general picture of the climatic and soil conditions, bearing in mind also that such conditions have considerable effect upon human activities.

The climate of the Peak District differs from that of the flanking lowlands in two main respects. Firstly, the effect of elevation is to reduce the mean temperatures and to increase the amount of precipitation. This, of course, is most strikingly illustrated by the conditions found on the high northern moors which reach to 2,000 feet in altitude. Here the mean temperatures for January and July are only about 34° F. and 52° F. respectively, while the rainfall exceeds 60 inches. Secondly, the diversified relief gives rise to marked local variations of climate, not only between the valleys and adjacent uplands but between one valley and another. In fact, generalisations for the region as a whole are of little use, except that conditions are distinctly more severe than those in the lowlands.

TEMPERATURE

Full long-term records are available only for Buxton, where the meteorological station stands at 1,007 feet above sea-level. Much of the moorland, however, lies considerably above this altitude, while many of the valleys reach much lower. At Buxton (see table) mean monthly temperatures range from 36·1° F. in January to 57·7° F. in July, with only June, July, August and September having means above 50° F. Comparative figures for Belper in the lower Derwent Valley at 280 feet are: January 38·2° F. and July 60·1° F. Although in general temperatures diminish with increasing altitude, notable exceptions frequently

CLIMATIC SUMMARY FOR BUXTON
Lat. 53° 15′ N., Long. 1° 55′ W., Altitude 1,007 ft.

	Jan	Feb	Mar	Apr	May	Jun	Jul	Aug	Sep	Oct	Nov	Dec	Year
TEMPERATURE (°F.)													
Mean	36·1	36·2	39·2	43·3	49·7	54·2	57·7	56·9	53·0	46·9	40·9	37·4	46
PRECIPITATION													
Mean (inches)	4·5	3·8	4·1	2·9	3·1	3·2	3·9	4·4	3·2	4·9	4·7	5·7	48·4
Rain days	19	18	19	16	16	14	16	19	15	19	19	21	211
Days with snow	8	8	8	3	1	0	0	0	0	1	3	6	38
SUNSHINE													
Hours per day	1·0	1·7	2·9	4·2	5·5	5·9	5·3	4·7	4·1	2·6	1·4	0·7	3·3
Percentage of possible total	12	17	24	30	34	35	32	32	32	25	17	9	27

Figures (based on Air Ministry records) given by E. G. Bilham in *The Climate of the British Isles*, p. 327

occur. Under anti-cyclonic conditions cold air at night gravitates from the plateau surface into the deeper valleys, reducing the temperature at the floor to as much as ten or more degrees below that of the slopes. Such temperature inversions may cause severe frosts in Edale, the Hope and other valleys, while their upper slopes and plateaux remain immune. An equally common feature occurring both morning and evening, is for the valleys to be filled with mist while the moors above are bathed in sunlight. Dr Alice Garnett has shown that inversions in the upland valleys last for twelve hours or more during late spring and summer, but with the onset of winter they may persist both day and night, sometimes for several days, depending on the continuance of anti-cyclonic conditions. Dr Garnett has also shown that the intensity of inversions varies with local physical factors. Thus Monsal Dale in the limestone area recorded a minimum temperature from 4° to 6° F. higher than those recorded on the same occasions in Edale, a valley in the gritstone uplands. Moreover, the minimum temperature in the water-filled upper Derwent may be several degrees higher than those of neighbouring valleys which have no reservoirs.

Frost is liable to occur on all the high ground, especially above 1,000 feet, between September and June. At Buxton itself, ground frosts are experienced on an average of 111 days in the year, the month of greatest frequency being March.

PRECIPITATION

Heavy precipitation is characteristic of most of the Peak District, partly on account of the considerable elevation and partly because, like the southern Pennines as a whole, the region lies directly in the track of depressions moving eastwards from the Atlantic. Moreover, the comparative frequency of polar maritime air in winter results in a good deal of snow on the higher parts in addition to the abundant rainfall. The mean annual precipitation diminishes broadly in relation to altitude from over 60 inches on the lofty summits of Kinderscout and Bleaklow (2,000 feet) in the north-west to under 35 inches in the valleys of the south and south-east such as the lower Wye, the Derwent (between Baslow and Great Rowsley) and the Bradbourn valley (Fig. 7). Buxton, at just over 1,000 feet, receives 48·4 inches, while Chatsworth in the Derwent Valley at 436 feet, less than 12 miles to the east, has only 34·7 inches. Yet there are sharp differences between the amounts of rainfall on relatively high and exposed surfaces and those in sheltered valleys close at hand. The highest recorded figures are those of 63 inches on the northern edge of the Kinderscout plateau and on Bleaklow, both districts reaching to over 2,000 feet. Over the high ground to the south-west of Buxton, including Axe Edge, the mean annual amount exceeds 50 inches. In fact, almost all the moorland in the western half of the region from Saddleworth Moor in the extreme north to Warslow in the south receives more than 45 inches.

Some idea of the incidence of rainfall through the year in The Peak may be obtained from the records at Buxton (1928-54). Here March and May are the only months with less than three inches, although April has but slightly more than this amount; while each of the four months October-January inclusive has over five inches, their combined amount being nearly 45 per cent of the yearly total. Even the three summer months May, June and July between them receive almost a quarter of the annual rainfall. The number of rain days in the year at Buxton (i.e. days on which 0·01 inch or more falls) is 211, ranging from 21 in December to 14 in June. Belper, however, in the lower Derwent Valley with a mean annual rainfall of just under 31 inches, has only 180 rain days.

Snow is recorded at Buxton on an average for 38 days in the year. It occurs on eight days in each of the winter months January, February and March, and on six days in December. On higher ground snow is much more frequent and persists longer. At times the main roads across The Peak from Sheffield to Manchester, as well as those leading out of Buxton, are liable to be blocked. Few winters pass without one or other of these routes being rendered impassable, usually on account of drifting snow, for periods varying from a few hours to several days. In these parts the large snow-plough lying in readiness at the roadside is a common feature.

SUNSHINE

For sunshine the Peak District is a rather disappointing region. Low cloud, fog and mist, all largely resulting from the elevation, combine to reduce sunshine to small amounts, especially in winter. At Buxton the average daily amount through the year is only 3·3 hours or 27 per cent of the possible amount. In winter the figures are particularly low, being 0·7 hour or 9 per cent in December and 1 hour or 12 per cent in January.

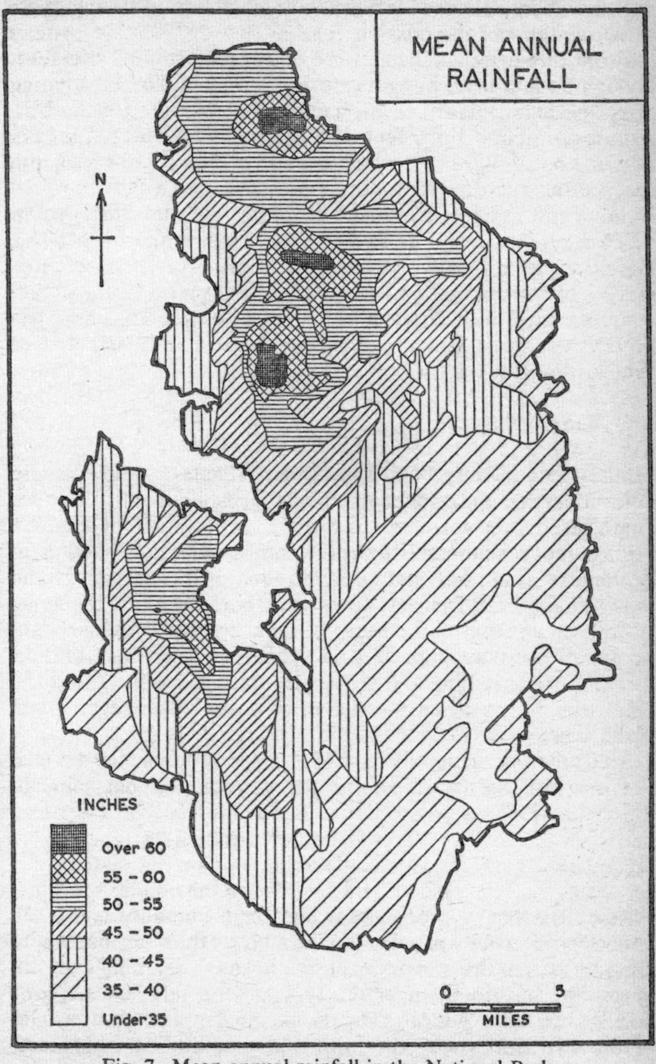

Fig. 7 Mean annual rainfall in the National Park

The daily average, however, exceeds 5 hours during the summer months (May, June and July), but even this is barely more than one-third of the possible total.

From this brief analysis of the leading aspects of climate it is obvious that in some respects conditions in The Peak verge upon the harsh. Truly severe conditions characterise the high moorlands above 1,500 feet, where low cloud, hill fog, heavy rainfall and snow, strong winds and comparatively low temperatures, often very low in winter, conspire to give a climate which is not even mitigated by generous hours of sunshine. Below 1,200 feet, however, conditions are ameliorated, vegetation flourishes more readily, cultivation is more practicable and less risky and human settlement is accordingly more widespread, adapting itself to a climate which may be described as intermediate between that prevailing in the lowlands and that of the high moors.

SOILS

While it has already been stated that climate and soil largely determine the appearance of the landscape, it must also be emphasised that soil itself is largely the product of climate. Soils in fact result mainly from the combined effects of climate, geological conditions and surface relief. In a relatively small area like the Peak District, however, where the climatic régime is broadly uniform, differences of rock type and configuration may well be more important in promoting variations of soil. Actually the gritstone and grit shales on the one hand and the limestone on the other afford marked contrasts as regards soil conditions.

The gritstone generally provides a sandy loam, the texture varying with the nature of the parent rock, so that some of the coarse-grained grits produce a gravelly soil. On the damp high moors, however, a peat bog of the blanket type covers much of the surface. This tends to be deep on flat plateau surfaces such as Kinderscout and shallow on the convex marginal slopes. The heavy precipitation and high humidity cause the deeper peats to be water-logged throughout the year, but where the peat is thin the moisture causes thorough leaching and promotes the development of the typical lime-deficient ash-grey soil known as a *podsol*. The Edale Shales and other shales of the Grit Series, together with those of the Coal Measures forming part of the East Moor, are all somewhat similar in their physical and chemical characteristics and give a cold and rather sour clayey soil generally grey or yellow-grey in colour.

The soils of the Carboniferous Limestone are of a different character. Although the limestone area is of a lower mean elevation than the gritstone, most of it nevertheless lies at a height of over 900 feet so that cool humid conditions still prevail. Both surface and subsoil, however, are freely drained and waterlogging occurs only on the gentler slopes during winter and spring. The characteristic soil is brown in colour and of medium texture; it is readily workable despite the fact that the subsoil is generally rather stony or gravelly. In detail, however, the limestone soils vary considerably with the surface relief. On comparatively steep slopes only a thin soil develops, usually under two inches, while rapid drainage prevents leaching. This results in a type of soil known as *rendzina,* which has an appreciable lime content derived from the parent rock. On more moderate slopes the *rendzina* is deeper, drainage is slower and some leaching takes place, in consequence of which the lime content is lower. In some parts this process is carried further so that the removal of lime, and the accumulation of a non-calcareous clayey residue as a result of leaching, produce a soil of the *brown earth* type which may develop to a depth of 2-3 feet. On the flat upper surfaces above 900 feet, where the soil is even deeper, increased moisture favours leaching, which in turn gives rise to the formation of what may be termed an incipient *podsol.* In such areas lime-loving (or calcicole) and lime-avoiding (or calcifuge) species can sometimes be seen growing side by side. This phenomenon arises from the differing lengths of the rooting systems. Plants which have only short roots must of necessity grow in the surface layers of the soil which are lacking in mineral salts and are distinctly acid, whereas the longer-rooted species extend their rooting systems down to the calcareous and alkaline layers below. Examples of this association of calcicole and calcifuge plants can be seen on Wardlow Hay Cop and Longstone Edge, where there is a certain amount of limestone heath. On a few heights where a capping of chert (a siliceous deposit locally accompanying the limestone) forms the surface, as on Bradwell Moor (1,400 ft.) and to a lesser extent on the southern slope of Miller's Dale, true *podsols* are found.

A further complication results from the numerous igneous intrusions which occur throughout the limestone country. These are often small in area, as in Miller's Dale and Tideswell Dale, but they produce soils of a different character with a consequent local modification in the flora, and it is not uncommon to see calcifuge species like bilberry *(Vaccinium myrtillus)* forming a colony surrounded by characteristic calcicolous plants such as salad burnet *(Poterium sanguisorba)*.

Soils derived from river alluvium are of little significance in the Peak District for the streams are too swift and their valleys too narrow for the development of flood-plains. As might be expected, the largest areas of alluvium occur within the outcrop of the shales, such as that deposited by the River Noe in lower Edale and the Hope Valley and that by the Wye near its confluence with the Derwent at Rowsley. Composed of uniformly fine particles, the soils of these areas are generally heavy and as a rule dark in colour.

From the standpoint of their value to agriculture the soils of the limestone are much superior to those of the gritstone both for cultivation and pasture. This is chiefly because they can be more easily worked and are naturally better drained. To a greater or lesser degree the soils of both formations are lime-deficient, so that even on the limestone applications of lime are necessary to make the land fully productive. At the other extreme, the least productive ground of all, on which few plants of any kind will grow, is to be found in the cloughs and on the more rugged flanks of Kinderscout and Bleaklow. In such places the continual exposure to weathering, either through frost-splintering or the effect of running water, prevents the formation of true soils. All that results is an accumulation of coarse rock-waste and sand, sometimes called skeletal soil, and even this is liable to be removed by swollen torrents.

REFERENCES

Manley, G. Snow Cover in the British Isles. *Met. Mag.: 76* (1947)
Manley, G. *Climate and the British Scene.* Collins: New Naturalist (1952)
Garnett, Alice. Section 3, Climate from *Sheffield and its Region.* British Association for the Advancement of Science (1956)
Pigott, C. D. Soil formation and development on the Carboniferous Limestone of Derbyshire. *Journ. Ecol.: 50* (1962)
Bryan, R. Climosequences of Soil Development in the Peak District of Derbyshire. *East Midland Geographer: 28* (1967)

5. The Forest

> And he spake of trees, from the cedar tree that is in Lebanon even unto the hyssop that springeth out of the wall
>
> *I Kings iv, 33*

The story of the evolution of the Peak District scenery told in the earlier chapters explains how the present diversity of relief, ranging from the deep dales to the high plateaux, has come about. Equally important in accounting for the appearance of the landscape is the mantle of vegetation which in one form or another covers almost all the ground. For a full appreciation of the scenery, therefore, some attention must be given to the character of the vegetation.

Much has already been published on the flora of The Peak. The new *Flora of Derbyshire* edited by A. R. Clapham which appeared in 1969 provides the most up-to-date and comprehensive account, while Linton's *Flora* of 1903 is still a useful work of reference. A modern *Flora of Cheshire* by Newton was published in 1971, and a *Flora of Staffordshire* by E. S. Edees is expected shortly. These floras cover areas where the Peak District extends into other counties, as do such older works as *Flora of West Yorkshire* (Lees, 1888), the *Flora of Cheshire* (de Tabley, 1899), and the *Flora of North Staffordshire* (Boyden Ridge). Floras, however, are concerned with little more than recording the occurrence and distribution of individual species and take little account of the relationships of those species one with another and with their environment. Examination of the plant life of unenclosed and uncultivated lands such as can be found extensively in The Peak shows that the individual species are grouped into plant communities or vegetation units. Study of these constitutes the modern ecological approach to vegetation, and Dr C. E. Moss's *Vegetation of the Peak District* (1913) represents one of the pioneer works in this field.

The types of vegetation found within the Peak National Park provide an admirable opportunity for studying the relationship between plant communities and their habitats. Moreover, since The Peak forms part of the southern Pennines and is at the same time penetrated by valleys leading from the bordering lowlands, especially from the south, it is a region in which northern and southern types of British vegetation meet and overlap and where many species attain their southern or northern limits.

In the broadest sense the character of vegetation over large areas is determined by climate. In Britain, however, as in much of north-western Europe, there has been a succession of climatic changes during the relatively short period of post-glacial time, so that corresponding changes in vegetation have occurred. To some extent, nevertheless, the present vegetation bears resemblances to that of preceding phases, for the changes in climate were not extreme.

DEVELOPMENT OF THE FOREST

Recent research has thrown much light on the previous forms of vegetation and as far as the Peak District is concerned a fascinating guide to past changes is afforded by the technique of pollen analysis. Dr V. M. Conway's investigation of the Ringinglow peat bog by this method provides a record which is representative of the high moorlands. It shows that the tundra vegetation existing on the ice-free uplands during late-glacial times was followed as the climate became warmer (Boreal period) by birch forest in which Scots pine gradually became abundant. About 6000 BC, under damper conditions (Boreal-Atlantic transition period), peat bog began to form on the uplands including the shallow basins of the moor where Ringinglow is situated. The approximate date of this development is confirmed by the finding of late Mesolithic implements beneath the peat. For a long time peat formation was slow, as is shown by the discovery of Bronze Age implements only a few inches above Neolithic flints. About 1200 BC, however, wetter conditions began to prevail (Sub-Atlantic period) and peat accumulation became more rapid. Naturally as the area of peat extended, the upper limit of the woodland was lowered. Then about 500 BC came a further acceleration in peat bog development with a rapid advance of sphagnum moss. It was about this time, i.e. some 2,500 years ago, that the conditions of the cool temperate climate as we experience them today set in.

Now if Nature were left to herself, unhampered by man's interference, the characteristic form of vegetation as developed under our present climate would be mixed forest. Woods of oak, ash and other deciduous trees together with some conifers would be found over all the lowlands, while on much of the higher ground oak and ash would persist, along with birch and probably Scots pine. Higher still, under exposed conditions, the trees would become smaller and even stunted, while in the highest parts of our mountain areas conditions would favour such plants as heather and bilberry rather than tree growth. There would of course be some differences in the predominance or otherwise of particular trees according to soil and drainage conditions. For thousands of years, however, stretching back to well before the climate assumed its present form, our vegetation has not been left to develop naturally, except in a very few places. Human activity in one form or another has constantly interfered with the plant life, causing changes and modification. Today, therefore, in most parts, even where it gives every appearance of being natural, the vegetation is really only semi-natural, and this is as true of the Peak District as of the rest of the country. Despite this, it is important to realise that the existing vegetation still shows a close relationship to soil and other physical conditions.

THE REMOVAL OF THE FOREST

Of course the most obvious change which man has brought about during his occupation has been the removal of an enormous proportion of the forest which once covered the greater part of the land. The chief reason for doing so was to bring the surface into agricultural use either for grazing or cultivation. This process of deforestation took place in stages both in prehistoric and historic times, more rapidly at some periods than at others and more completely in some areas than in others.

We may now glance at some of the ways in which human activity has contributed to the removal of the forest cover in The Peak. First, it should be noted that it is only on the high northern moors, in the region of desolate bogs, that man's interference has had little or no effect. Even in these remote tracts, however, sheep-grazing has at least played a minor part in bringing about the withdrawal of the woodland from the highest altitudes, though the more important factors have been of a physical nature, the spread of peat and the progressive souring of the land. The combined result has been a lowering of the tree limit by several hundred feet within historic times, the limits of the oak and ash now being about 1,000 feet above sea-level and that of the birch about 1,250 feet. This, incidentally, has had the effect of rendering re-afforestation by deciduous trees almost impracticable today.

In all other parts of the region forest reduction has been so complete as to leave few traces, if any, of the original woodland. Padley Wood on the gritstone slope overlooking the upper Derwent is regarded by some ecologists as a surviving fragment of natural oakwood which once clothed the Burbage Valley and much more of the neighbourhood. In the limestone upland the ashwoods of Dovedale and some of the other dales are possibly relics of the natural climax woodland of that part of the region, though some authorities regard even these as being semi-natural.

From archaeological evidence it may be inferred that forest clearance for agricultural purposes began in either late Neolithic or Bronze Age times. At first the trees were cleared to provide small patches of ground for cultivation, and only much later on when the means of felling were improved did the process become systematic. In this connection evidence obtained from the pollen in the Ringinglow peat is highly suggestive. The pollen analysis showed that weeds of cultivation existed in both Neolithic and Bronze Age times. Since Ringinglow Bog is situated immediately to the north-east of the limestone upland, it is possible that this pollen may have been carried from the latter area to the bog by the prevailing south-westerly winds. This would tend to confirm the view, which is supported by other evidence, that clearance of the land for cultivation first occurred on the limestone. It is likely, however, that the removal of trees was begun even earlier by Neolithic man to extend his pastoral activities and it is probable that the regeneration of the woodland was in the first instance prevented by the grazing of his cattle.

During historic times the destruction of the forest was rapidly extended. Although the Roman occupation of the Peak District does not appear to have accelerated the process, except presumably in the places where lead was worked, it was the period of Anglo-Saxon colonisation which witnessed the most thorough and most widespread conversion of forest into farmland. The frequency of place-names containing the elements *-clough, -den* (=dean), *-hey, -hirst, -ley* (from lea), *shaw* and *withen,* testifies to the extent of the former woodland. Here a little more light may be shed on the question of the date of clearance in the limestone upland. Not many of the Saxon names reflecting forest clearance occur in that area, although some containing the element *-ash* are found, such as Ashford and Monyash. This may indicate that the removal of the woodland was already well advanced by Saxon times. Further support for this view may be gleaned from the fact that a century or two later the limestone portion of Peak Forest became known as 'campana', a term allied to the French *campagne* referring to open country. Thus on various grounds it may be adduced that the original forest on the limestone of the Low Peak was removed during the earlier phases of human settlement from all but the very highest parts and some of the more inaccessible dales in which relict ashwoods still remain.

Under the Norman kings, a set-back to agriculture was caused by the creation of the Royal Forests to serve as hunting grounds. These areas were placed under rigorous Forest Laws which forbade cultivation. Peak Forest and Macclesfield Forest date from this time. Like most of the others they were largely waste and not heavily wooded, but they were mainly based on natural vegetation. Peak Forest formed a large tract which was kept as a reserve for the hunting of boar and red deer. The name of Chapel-en-le-Frith (Frith = Forest), a settlement founded by the King's foresters in 1225, is an interesting reminder of this period. In other parts of The Peak throughout the Middle Ages the demand for timber and brushwood for use in the mining and smelting of lead caused further loss to the tree cover.

Dr S. R. Eyre in a detailed study of the East Moor has dealt convincingly with the effect of human action upon the woodland. According to the Domesday Survey of 1086 there were 38,766 acres of forest on this upland, but at the time of the Parliamentary Enclosures in the eighteenth century, there were only 2,470 acres. Pollen analysis confirms that there was a steady reduction in the amount of tree pollen from the twelfth century onwards until it disappeared almost completely in the seventeenth century. Documentary evidence shows that not only were timber and brushwood used for many purposes, but that grazing was an important factor in preventing natural regeneration. By the end of the thirteenth century, for example, the monks of Beauchief Abbey near Sheffield and several lords of the Manor in the neighbourhood held grazing rights over large tracts of the Moor. Similar circumstances undoubtedly prevailed over other parts of The Peak.

THE RESTORATION OF WOODLANDS

Now there are limits to the reduction of the natural forest cover beyond which it is reckless and foolish to proceed. In modern times Man has come to realise this and in the Peak District, as elsewhere, some measures have been taken to restore the woodland. This has been done for various purposes, for the provision of timber, for the protection of game, for the regulation and protection of water supply and not least, in the case of large estates, to add to the amenities of the countryside. These attempts to reverse the centuries-old trend have in some places ensured the preservation of ancient woods, while in others, replanting with different species has produced a significant revival. Thus Padley Wood, mentioned previously as an example of ancient oakwood which may be of natural origin, has been carefully maintained and may still be regarded as self-generative. Old woods of mixed oak and birch remain on the slopes of Longdendale, while on the other hand the oakwoods of the Woodlands Valley have been replaced by coniferous plantations. From a study of late eighteenth-century Enclosure awards affecting parts of the East Moor it would appear that the woods, chiefly of oak, on Burbage Rocks, Blackstone Edge and Gardom Edge have been appreciably enlarged.

It was the landed gentry of the eighteenth century in particular who made a splendid contribution to the recovery of the woodland. To their deliberate designing of the landscape we owe much of the present-day appearance of the countryside. They reclothed many of the slopes (which in time yielded valuable timber) and graced their parks and farmland with noble trees. Instances of these landscape improvements are many. From 1771 onwards Sir Richard Arkwright planted 50,000 trees annually along the Derwent Valley above and below his mills at Cromford.

Nearby, the Gells of Hopton added considerably to the trees on the slopes of the Via Gellia when they built the road along the valley in the early nineteenth century. The outstanding examples of this form of landscape development, of course, were those promoted by the Dukes of Rutland and Devonshire around their great mansions at Haddon Hall and Chatsworth respectively. At Chatsworth the grounds around the house were originally laid out by George Loudon about 1690 when the rebuilding by the first Duke took place. Thereafter, as the estates

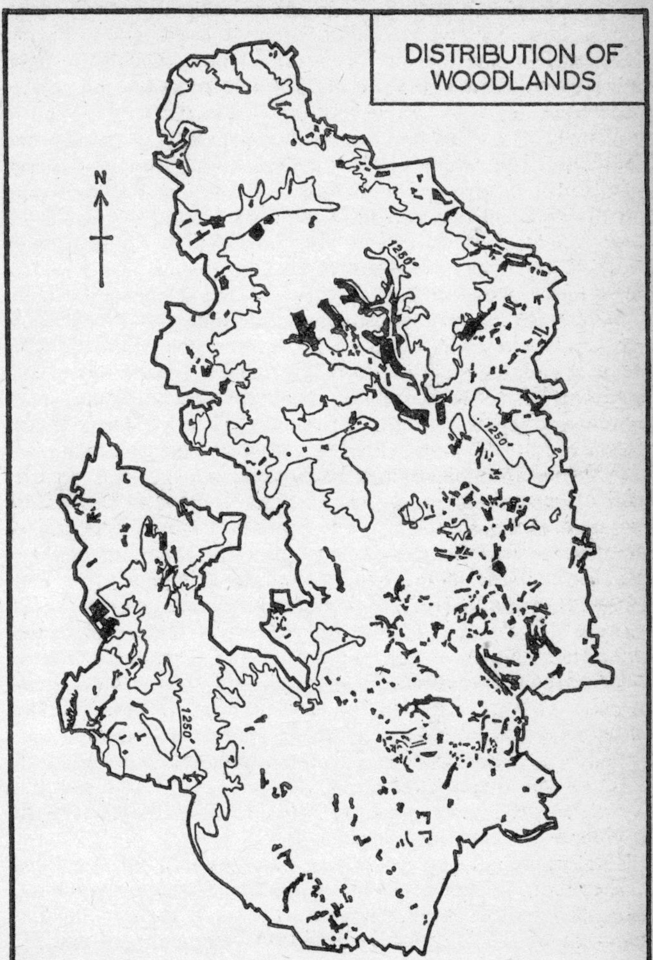

Fig. 8 Distribution of woodlands in the National Park. Note the almost complete absence of woodland above 1,250 ft.

were enlarged, many areas within the park and beyond were planted, including the woods in the neighbourhood of Beeley and Baslow. Several famous landscape planners took part in the embellishment of the Devonshire lands, though the most lavish expenditure was naturally devoted to the formal gardens and park. Capability Brown was consulted in 1760 by the fourth Duke, and Joseph Paxton directed further improvements in the mid-nineteenth century under the sixth Duke. Similar developments were inspired from Haddon Hall and much of the woodland established between 1800 and 1825 around the Hall itself and along the beautiful stretch of the Wye Valley from Bakewell to Taddington, are the result of the liberal aims and foresight of the fifth Duke of Rutland and his wife, the Duchess Elizabeth.

Nineteenth-century development for amenity is well illustrated at Buxton, where many visitors today appreciate the leafy shade afforded by Corbar Woods which were planted over disused quarries to the north-west of the town, while to the south the canopy of trees on the slopes of Grin Low help to make the ascent to Solomon's Temple a cool and pleasant walk.

From the foregoing it will be evident that the Peak District today is not a heavily wooded region. Neither is Derbyshire as a whole, for the Forestry Commission's census of 1947-9 showed that woodland occupied little more than four per cent of the total area of the county, compared with six per cent for England and Wales. The woodland area amounted to 26,000 acres in all, of which about 15,000 acres were under high forest, i.e. stands of trees which had grown to full maturity. Nearly two-thirds of the latter consisted of deciduous woods, about one-third of coniferous woods and plantations, while some 2,000 acres ranked as mixed woodland. In view of the modern trend in timber demands, the acreage of coniferous trees is expanding more rapidly than that of the hardwoods.

Although the woodlands tend to be scattered and are generally lacking on the high moors; trees are nevertheless abundant in the Peak District and many fine individual woods are to be found (Fig. 8). Trees are certainly one of the primary attractions of the National Park, and their colour, changing with the seasons, gives it a loveliness not excelled by any other part of the country. Apart from the fact that the highest gritstone moors lie above the climatic limit of most deciduous trees, the present pattern of distribution is largely the result of early clearance for agricultural purposes, especially on the limestone plateau, and of re-planting undertaken at various times in the more recent phases of human history. Even on the limestone, despite its use for crops and grass, trees are often abundant; in fact this landscape of small fields bounded by the rather sombre stone walls is pleasantly enlivened by clusters of deciduous trees which take various forms ranging from small rectangular plantations to little copses screening the farmsteads or isolated clumps standing like sentinels on the hill-tops.

The chief deciduous trees are the oak, ash, sycamore and beech. Of these the sycamore is the most extensive, covering 3,500 acres of forest, with an additional 1,000 acres in coppice, but the woods are neither so large nor the timber so valuable as they are farther north in Yorkshire. The ash, which occupies some 1,300 acres of high forest and an almost similar amount of coppice, chiefly in the limestone dales, is now in danger of neglect, though not so long ago, on account of the toughness and resilience of its timber, it was of great value to Sheffield toolmakers in providing wood for handles. The most important coniferous tree is the Scots pine, which covers half the acreage under softwood timber. A good deal of Sitka spruce has been established around the Derwent reservoirs, while in the so-called Hope Forest (Alport and Ashop valleys) Japanese larch has proved successful.

Besides their amenity value, woodlands have, of course, several other functions in the Peak District. Many small woods and tree-belts are maintained to give shelter to crops, cattle and game, while as a source of timber all the sizeable woods represent, at least potentially, an important economic asset. Since most of them are privately owned it is unfortunate that estate forestry has sadly declined in recent years. There are notable exceptions such as the Devonshire and Rutland estates, where the woods are still kept under careful management and yield a steady production of timber. Here, however, a relative disadvantage of location arises. It is found that polluted air from the industrial areas such as Sheffield and north-east Derbyshire on one side and Manchester and south Lancashire on the other produces some ill-effects on the growth of hardwood trees. Not only do such trees have a dull and occasionally a blackened appearance as the result of sootladen air, in contrast to the fresher green of similar species growing in places remote from industrial centres, but they require much longer to reach maturity. Observation made a few years ago showed that trees belonging to the Chatsworth Estate were taking almost twice as long to mature as similar hardwoods in southern England.

This was indicated by the shortness of the internodes on twigs and branches. In the past decade or so there has been so much cleansing of city air that pollution is noticeably less, and in time this should influence tree growth and appearance.

Apart from the few large estates the most productive woodlands at the present time are the few hundred acres owned by the Forestry Commission and the coniferous plantations established by the water-supplying authorities. The Peak Park Planning Board plants about 20,000 trees each year, uses Tree Preservation Orders to safeguard important woodlands, and manages, in all, a total of 350 acres of forest.

REFERENCES

Moss, C. E. *Vegetation of the Peak District.* Cambridge University Press (1913)

Woodhead, T. W. History of the Vegetation of the Southern Pennines. *Jour. Ecology: 17* (1929)

Conway, V. M. Ringinglow Bog near Sheffield. *Jour. Ecology: 34 and 37* (1947 and 1949)

Forestry Commission. *Census of Woodlands: 1947-49.* H.M.S.O. (1952)

Edlin, H. L. *England's Forests.* Faber and Faber (1958)

Scurfield, G. The Ashwoods of the Derbyshire Coniferous Limestone: Monks Dale. *Jour. Ecology: 47* (1959)

Merton, L. F. H. The History and Status of the Woodlands of the Derbyshire Limestone. *Jour. Ecology: 58* (1971)

6. The Major Plant Associations

What would the world be, once bereft
Of wet and of wildness? Let them be left,
O let them be left, wildness and wet;
Long live the weeds and wilderness yet.
Gerard Manley Hopkins

In addition to what remains of the woodland cover, the plant mantle of the Peak District includes a wide variety of ground vegetation, and it is this which should now be considered. As already stated there is a close relationship between the plant communities and soil conditions. This means that not only is there a difference in the broad vegetation types between the gritstone and limestone areas of the region, but that in detail the plant communities differ in respect of the types of soil found on each of these rock formations.

In the gritstone area, the lower parts of the moorland (including the outcrops of the Edale Shales and other grit-shales) support a grass cover on the damper acid soils in which Mat Grass *(Nardus stricta)* is dominant. On the drier soils *Agrostis tenuis* and *Festuca ovina* are abundant. This grass cover occurs chiefly on slopes of 1,000-1,250 feet, and is termed siliceous grassland. It is usually succeeded by a Heather *(Calluna vulgaris)* zone up to about 1,500 feet, above which the heather gives way to Cotton Grass *(Eriophorum vaginatum)*. The last-named, which is botanically one of the sedges and not a grass, is characteristic of the high northern moors and forms one of the most extensive tracts of such vegetation to be found in the Pennines. It is well established on the peat, and thus forms the main plant cover on some of the high moors. These moors, which are the haunt of the curlew, are wild and desolate for much of the year, but in June and July are strikingly transformed by the white mass of Cotton Grass. On the West Moor, with its prevailing steep slopes, the Cotton Grass is confined to the more level stretches around the head of the Goyt Valley and that of the Dane. The East Moor, although comparatively low in altitude, supports small areas of Cotton Grass such as can be seen on Totley Moss and Leash Fen.

Although the Cotton Grass zone has proved of little utility to man, it has nevertheless suffered from some interference, as has the Heather zone. The latter has increased appreciably in recent times, especially on the grouse moors. The practice of burning, as part of grouse moor management since the mid-nineteenth century, and even earlier for sheep-grazing, has greatly assisted this extension. In this respect at least part of the existing Heather zone can be regarded as the product of human interference.

Over the limestone area soil is the leading factor in determining the characteristic vegetation. As shown in chapter 4, the chief types of soil show a range from immature *rendzina* to *brown earth* and *podsol*. Since these stages are largely the product of slope, the resulting vegetation types also show a broad zonal arrangement. The steep dale sides with their immature *rendzinas* support an ashwood community. Even grazing on some of these slopes is not effective in preventing natural regeneration, while some small areas of Ash, protected from livestock by cliff or scree, may be regarded as truly primitive woodland. Ashwoods are found in all the major dales. In Dovedale, looking from Lover's Leap towards the Staffordshire side, a succession of young, mature and decaying trees can be seen, showing continual regeneration. The journey by train through Miller's Dale in early summer affords a vista of Ash foliage at its loveliest.

On the plateau edge, the soil changes and the ashwoods tend to degenerate into scrub with Birch *(Betula pendula)* sometimes replacing the Ash. Both Hawthorn *(Crataegus monogyna)* and Hazel *(Corylus avellana)* are characteristic species of this scrub with a wide variety of herbaceous plants forming a ground layer beneath them. Most of the plants in this community are typical calcicolous species in consequence of which the vegetation is termed calcareous scrub.

On the gentler and more stable slopes of the dales, as well as on the plateau, the more mature *rendzina* soils support a calcareous grassland which is dominated by Sheep's Fescue *(Festuca ovina)*. Associated with stream courses is a special type of vegetation rich in species adapted to water and to the damp conditions along the banks.

On the plateau proper, the brown earth soils form the most suitable ground for agriculture. This is devoted overwhelmingly to pasture with only a few arable fields. The 'improved' grassland, used mainly for dairy cattle, is maintained only by constant grazing and manuring, for wherever fields are neglected a calcifuge community develops in which Bent Grass *(Agrostis)* is the main constituent. On the highest parts of the plateau where soils verge towards the *podsol* type the grassland contains Mat Grass and Waved Hair-Grass *(Deschampsia flexuosa)*, with Bilberry and Heather. On Bradwell Moor, with its stretches of peat, true *podsols*, pure Heather and Bilberry areas occur.

From this very general analysis, the broad types of vegetation found in the Peak District may be summarised as follows: firstly, in the gritstone and shale areas the siliceous grassland with its particular plant communities must be distinguished from those of the high moors, which in turn exhibit broad zonal distribution in relation to altitude as well as soil conditions; secondly, in the limestone country the distinctive types of vegetation consist of the relict ashwoods, calcareous scrub, calcareous grassland, and the plants associated with the limestone streams. On this basis the different forms of the vegetation cover will now be examined in detail.

PRESENT-DAY PLANT COMMUNITIES

As already seen, the close correlation between soil characteristics and the plant cover has resulted in two broad types of vegetation in the Peak District, that associated with the siliceous soils produced on the sandstones and shales, and that associated with the calcareous soils of the limestone. At the present day the siliceous soils, particularly on the gentler slopes, are for the most part clothed with coarse grass which is widely used for pasturing sheep, but where the grazing element is removed this grassland is soon invaded by Heather *(Calluna vulgaris)*, Gorse *(Ulex europaeus* and *U. gallii)*, and Bracken *(Pteridium aquilinum)*. This invasion by heathy species in the absence of grazing is even more marked at higher altitudes, while in some of the cloughs there is still a certain amount of woodland dominated by the Sessile Oak *(Quercus petraea)* in association with the two Birches *(Betula pubescens* and *B. pendula)* and the Mountain Ash *(Sorbus aucuparia)*. This latter community constitutes the climax vegetation on these soils, though the examples of it are little more than fragmentary today. On the gritstone plateaux the siliceous grassland and the woodland fragments give place to moorland with extensive deposits of peat which may be six feet or more in depth. This is a habitat which supports communities of Heather, Bilberry *(Vaccinium myr-*

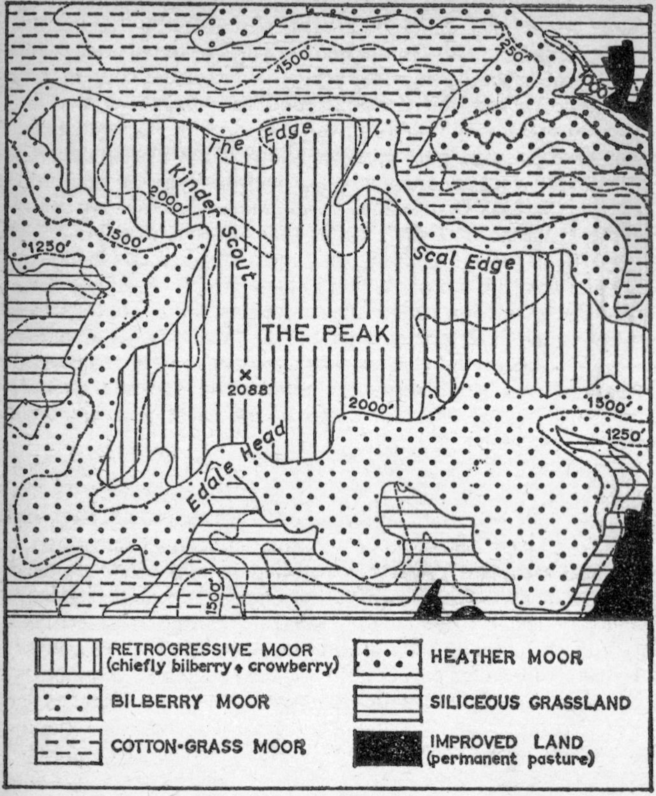

Fig. 9 Vegetation of the Kinderscout Plateau. Scale approximately one inch to one mile. (Based on C. E. Moss)

tillus), and Cotton Grass *(Eriophorum vaginatum* and *E. angustifolium)*, though on certain moorlands above 2,000 feet there are considerable areas which are almost completely denuded of vegetation. Such is the case in many parts of Kinderscout and Bleaklow. These provide examples of retrogressive moorland in which soil conditions are so excessively acid as to inhibit the growth of all but the most tolerant species. Thus the lack of plant cover and the slight convexity of the mountain-tops provide conditions which are ideal for the rapid and severe erosion of the peat which is carried in a semi-fluid state into the intricate network of drainage channels which dissect the plateaux (Fig. 9).

The limestone region is infinitely richer in species than the gritstone country, and several distinct communities can be recognised. The present day natural succession leads to Ash *(Fraxinus excelsior)* up to an altitude of 1,000 feet, but today only a few semi-natural ashwoods remain and these are confined to the slopes of the dales. The finest example is to be found in Hurt's Wood on the Staffordshire side of Dovedale, while another characteristic, though less extensive stretch occurs on the eastern side of Ravensdale. In quite a number of the dales the ashwoods merge almost imperceptibly into scrub in which Hawthorn *(Crataegus monogyna)* and Hazel *(Corylus avellana)* are often co-dominant species. In much the same way the scrub zone gradually gives place to calcareous grassland especially at higher altitudes.

The limestone uplands are much influenced by the activities of man, and in contrast to the corresponding regions on the siliceous soils they are divided into a vast number of relatively small fields by drystone walls which give the characteristic austerity to much of the Peakland landscape. These fields are used primarily for permanent pasture and meadows, with arable land found only here and there. Quite understandably the biotic influence in such areas has a profound effect upon the vegetation and strictly limits the number of species which occur. In addition to these three broad types of vegetation found on the limestone, some of the dale slopes support mixed plantations of marketable timber, and in one or two cases pure beechwoods can be found though these are in no way native to the district.

SILICEOUS GRASSLAND ON THE MILLSTONE GRIT AND EDALE SHALES

This type of grassland is particularly well developed on the shales and to a lesser degree on the gritstone, so that it occurs extensively in Longdendale, in the Ashop and Alport valleys, and in the cloughs of the Bleaklow-Kinder region, as well as on the slopes leading to the West Moors. The exact constituents of such grassland are very largely determined by the varying conditions of soil and soil moisture. Where the ground is well drained and the soil not excessively acid there is an abundance of Sheep's Fescue Grass *(Festuca ovina)*, which is usually associated with the Common Bent *(Agrostis tenuis)*. The frequency of these species is somewhat variable but generally speaking on the best-drained slopes the *Festuca* is the more abundant owing to the fact that it is better adapted to a xerophytic existence. The 'Bent-Fescue' community is an example of a biotic climax, for if the extensive grazing, especially by sheep, is withdrawn, a rapid invasion by heath plants like *Calluna* ensues, followed by colonisation by the two species of *Ulex*, and by *Pteridium*. Many of the siliceous slopes are damper, however, and here the Mat Grass, or White Bent *(Nardus stricta)* assumes dominance, and this type of grassland is undoubtedly the most common in a district where the rainfall is relatively high. Frequently the Waved Hair-Grass *(Deschampsia flexuosa)* is widely associated with the *Nardus*, and at lower levels much *Agrostis tenuis* occurs. As with the 'Bent-Fescue' community of the drier habitats, there is a tendency for the *Nardus* grassland to be invaded by heathy species whenever the grazing factor is removed. The two species of *Ulex* appear and *Pteridium* becomes increasingly common except in the wetter flushes on the hillsides in the vicinity of springs. Here the association is marked by patches of the common Rushes, *Juncus effusus* and *J. conglomeratus*, and to a lesser extent *J. articulatus*.

Generally speaking, the *Nardus* grassland extends to the edges of the plateaux and there, in many instances, merges directly with the heather moorlands which clothe the hill-tops. In the other cases where the slope is more gentle and conditions consequently wetter it gives place to grassland dominated by the Purple Moor Grass *(Molinia coerulea)*. Well-developed examples of *Molinia* grassland occur in parts of Big Moor, Barlow East Moor, and Brampton East Moor, and in The Peak it is invariably associated with ill-drained acid peaty soil and forms a link between siliceous grassland as a whole and true moorland.

The following list[1] of some of the characteristic plants of the siliceous grassland of The Peak gives an idea of the extent of the flora. A few plants common to the moorland habitats

[1] In this and subsequent lists of wild flowers, the flowers are grouped in families. The English names used are those adopted in *Collins Pocket Guide to Wild Flowers* by D. McClintock and R. S. R. Fitter, an invaluable handbook for identification in the field. The botanical nomenclature follows J. E. Dandy, *List of British Vascular Plants*, Brit. Museum, 1958.

will be noted in the list, thus emphasising the close relationship between the two communities.

Creeping Buttercup *(Ranunculus repens)*
Square St John's Wort *(Hypericum tetrapterum)*
Elegant St John's Wort *(H. pulchrum)*

Mountain Pansy *(Viola lutea)*
V. lutea var *amoena*

Knotted Pearlwort *(Sagina nodosa)*

Petty Whin *(Genista anglica)*
Broom *(Sarothamnus scoparius)*
Greater Birdsfoot Trefoil *(Lotus uliginosus)*

Tormentil *(Potentilla erecta)*
Trailing Tormentil *(P. anglica)*

Sheep's Sorrel *(Rumex acetosella)*
Common Sorrel *(R. acetosa)*

Cross-leaved Heath *(Erica tetralix)*
Bell Heather *(E. cinerea)*
Cowberry *(Vaccinium vitis-idaea)*
Bilberry *(V. myrtillus)*

Felwort *(Gentianella amarella)*

Foxglove *(Digitalis purpurea)*
Wood Speedwell *(Veronica montana)*
Birdseye Speedwell *(V. chamaedrys)*

Lousewort *(Pedicularis sylvatica)*
Eyebright *(Euphrasia nemorosa)*

Common Butterwort *(Pinguicula vulgaris)*

Wood-sage *(Teucrium scorodonia)*
Bugle *(Ajuga reptans)*

Sheepsbit *(Jasione montana)*

Heath Bedstraw *(Galium saxatile)*

Devilsbit Scabious *(Succisa pratensis)*

Common Ragwort *(Senecio jacobaea)*
Golden-rod *(Solidago virgaurea)*
Sneezewort *(Achillea ptarmica)*
Marsh Thistle *(Cirsium palustre)*
Leafy Hawkweed *(Hieracium umbellatum* and *H. sabaudum* agg.*)*

Heath Rush *(Juncus squarrosus)*
Toad Rush *(J. bufonius)*
Hard Rush *(J. inflexus)*
Soft Rush *(J. effusus)*
Compact Rush *(J. conglomeratus)*

Jointed Rush *(J. articulatus)*
Hairy Woodrush *(Luzula pilosa)*
Good Friday Grass *(L. campestris)*
Heath Woodrush *(L. multiflora)*

Bristle Club-rush *(Scirpus setaceus)*

Tawny Sedge *(Carex hostiana)*
Common Yellow Sedge *(C. demissa)*
Carnation Sedge *(C. panicea)*
Hairy Sedge *(C. hirta)*
Common Sedge *(C. nigra)*
Remote Sedge *(C. remota)*
White Sedge *(C. curta)*
Oval Sedge *(C. ovalis)*
Flea Sedge *(C. pulicaris)*

Red Fescue *(Festuca rubra)*
Tufted Hair-Grass *(Deschampsia caespitosa)*
White Bent *(Agrostis stolonifera)*

Some of the hillsides which are now clothed with siliceous grassland possess numerous relict forest species of which *Anemone nemorosa, Oxalis acetosella, Silene dioica, Corydalis claviculata, Geranium robertianum, Vicia sepium, Viola riviniana*, and *Endymion non-scriptus* are examples. These provide evidence of the former tree cover in which Sessile Oak *(Quercus petraea)* was dominant.

THE GRITSTONE MOORLANDS

The high moorlands of the Peak District are best developed on the flatter plateaux of the Kinderscout-Bleaklow area, while those to the west of Buxton are more limited in extent on account of the highly folded character of the gritstone much of which supports siliceous grassland rather than moorland. These moors vary in altitude between 1,500 and 2,000 feet, whereas the extensive East Moor barely exceeds 1,000 feet. Broadly speaking, three main types of vegetation can be observed. The first is heather moor, which usually covers the well-drained and gentle upper slopes leading on to the plateaux. It forms a transitional zone between the siliceous grassland and the second type, which is the Cotton Grass Moor. The latter assumes dominance where conditions are wetter and the peat cover thicker. The Heather zone occurs on the western and northern slopes of Kinder and Bleaklow, and in the upper Derwent Valley, while Cotton Grass, with *Eriophorum vaginatum* dominant, is very extensive on Featherbed Moss, on Bleaklow, and on the ridges above Alport Dale. Cotton Grass Moor also occurs to a limited degree on Combs Moss, and at the head of the Goyt Valley, and in spite of the low altitude, Leash Fen and Totley Moss furnish good examples on the East Moor.

The third vegetation type is dominated by Bilberry *(Vaccinium myrtillus)*, though this is less widespread than either Heather or Cotton Grass. Like the Heather it prefers somewhat drier conditions and in consequence thrives best where there is a moderate gradient and a thinner layer of peat. It forms an almost continuous fringe below the Kinder plateau and at lower levels is co-dominant with the Heather, but nowhere else is it quite so abundant. As already mentioned, the plateaux of Kinderscout and Bleaklow are examples of retrogressive Cotton Grass moorlands, where in some areas the peat has been almost completely denuded of vegetation, thus exposing it to rapid erosion. At the summit of Bleaklow the peat cover has indeed been completely removed and the ground colonised by Mat Grass *(Nardus)* in association with Bilberry and Heather.

In view of the retrogressive nature of the vegetation of these high moors it is not surprising that the flora is somewhat disappointing. The paucity of species on Kinderscout and Bleaklow is no doubt related to the continuance of the degeneration of the peat cover, but the apparent loss of quite a number of plants during the past fifty years may also be influenced by the increasing atmospheric pollution from the industrial regions of northern Cheshire and south-eastern Lancashire. The deleterious effect is especially noticeable on the cryptogamic plant life of such a locality. The lichen flora of Kinderscout, for instance, is particularly poor, for these plants are extremely sensitive to pollution, and there are similar evidences of a decline in the bryophyte flora since the late nineteenth century.

The degeneration of a Cotton Grass moorland, such as Kinderscout, results in the gradual drying of the peat with the resultant loss of the more hydrophilous species, and, where plant cover is not entirely absent, the introduction of species which are better adapted to drier conditions. Such plants as Bilberry, Cowberry *(Vaccinium vitis-idaea)* and Crowberry *(Empetrum nigrum)* are of this type; in structure they lack the aeration canals so advantageous to the Cotton Grass, whose roots are deeply embedded in the waterlogged peat, but being shallow-rooted can establish themselves satisfactorily on the drier surface layers. They form distinct local societies, in which the

Empetrum is usually the most plentiful, on both Kinder and Bleaklow, and often they are associated with an attractive bramble, the Cloudberry *(Rubus chamaemorus)*. The remaining flora of higher plants on the Peak plateau is extremely limited. A certain amount of *Calluna* occurs, as does the Cross-leaved Heath *(Erica tetralix)*; the Heath Bedstraw *(Galium saxatile)* and the Tormentil *(Potentilla erecta)* can be found occasionally, while near the source of the Derwent the rare Bearberry *(Arctostaphylos uva-ursi)* still exists. Certain species which might be expected to be widespread at such elevations are remarkable for their rarity or complete absence. The Club-mosses *(Lycopodia)* have not been recorded for many years, and the Round-leaved Sundew *(Drosera rotundifolia)* occurs only sparsely. One gratifying rediscovery of recent years in the upper Derwent Valley, though not exactly in a moorland habitat, is the very lovely Ivy-leaved Bellflower *(Wahlenbergia hederacea)*.

The flora of the West Moors is somewhat richer in species than that of the High Peak, but even here some plants are surprisingly rare. Very little Sundew, for instance, has been seen during the past fifty years although there are many wet bogs where one might expect it to grow luxuriantly. The Round-leaved Sundew was found in small quantity on Goyt's Moss during 1958. A few other choice plants do occur, of which Marsh Andromeda *(Andromeda polifolia)* is possibly the best. This wet moorland species is found in small quantity in two localities where it flowers sparsely in suitable seasons, but there are signs that it is decreasing, probably on account of the partial draining of the bogs by a network of channels which have been cut in recent years. One or two peat bogs of particular interest are to be found at lower levels below the West Moors, and these possess a much wider range of flowering plants. Goldsitch Moss, situated between Axe Edge and the Roaches, is a good example. In normal season the ground here is excessively wet and in the numerous bog pools there is an abundance of the handsome Bog Asphodel *(Narthecium ossifragum)*. Nearby, trailing over the pale Sphagnum moss, there is a rich growth of the Cranberry *(Vaccinium oxycoccos)*. Nowhere in the Peak District is this species more luxuriant, but by contrast it is difficult to explain the absence of any of the Sundews which are just as typical of such a bog.

A few miles away to the east, in the neighbourhood of Hollinsclough, another peat bog, Moss Carr, can be found. This, too, deserves mention from the point of view of its flora. It is a low waterlogged area situated between the valleys of the Dove and the Manifold, and is one of the few localities where the Bogbean *(Menyanthes trifoliata)* still flourishes in the Peak District. In association with the Bogbean there is quite an abundance of the Marsh Cinquefoil *(Potentilla palustris)*, while the Cranberry is present too, though not in the same quantity as on Goldsitch Moss. In a favourable season the Marsh Violet *(Viola palustris)* flowers luxuriantly, but probably the most striking plants here are the Marsh Orchids. The marsh form of the Spotted Orchid *(Dactylorhiza maculata* subsp. *ericetorum)* is widespread in association with the Dwarf Purple Orchid *(Dactylorhiza purpurella)*, which was discovered in 1946. As is common with the palmate orchids, these two hybridise freely on Moss Carr and the resultant cross, *Orchis* ×*formosa*, is frequently a more robust and striking plant than either of the parents.

The East Moor is probably the richest of the Peak moorland habitats from the point of view of variety of species, a feature which may be due, in part at least, to the much greater freedom from atmospheric pollution than is the case in the High Peak itself. As previously stated, these moors are dominated largely by heather communities, but have limited tracts of Bilberry and, in wetter habitats like Totley Moss and Leash Fen, there is much Cotton Grass. In quite a number of sites where the Bilberry and Cowberry occur, a more luxuriant growth compared with that in the Kinder region can be observed. In a similar manner the amount of fruit produced by these plants on the East Moors is decidedly greater. While this is probably due to the general habitat conditions being more favourable, it is also possible that there may be some relationship to the purer atmospheric conditions in this area. In considering members of the genus *Vaccinium* on the East Moor mention should be made of the hybrid between Bilberry *(Vaccinium myrtillus)* and the Cowberry *(Vaccinium vitis-idaea)*, which occurs in a number of localities from Hallam Moors southwards as far as the Matlock district. This hybrid, *Vaccinium* ×*intermedium* forms a robust plant often two or three feet in height, and bears quite large pale pink flowers which show obvious affinities to the two parents. Its special interest lies in the fact that it was first discovered on Cannock Chase in 1872, which site remained the only locality for it until it was found on the East Moor by the late Dr Eric Drabble in 1919. There are few other authentic records for it outside Staffordshire, Derbyshire and Yorkshire. A further species which deserves mention for the East Moor on account of its absence from other moorland tracts in The Peak is the Lesser Skullcap *(Scutellaria minor)*, which can be found in small quantity in several places.

The most characteristic plants of the Peak moorland habitats are included in the following list, though it cannot be considered an exhaustive one.

Bracken *(Pteridium aquilinum)*
Lady Fern *(Athyrium filix-femina)*
Lemon-scented Fern *(Thelypteris oreopteris)*
Beech Fern *(T. phegopteris)*

Lesser Spearwort *(Ranunculus flammula)*

Marsh Violet *(Viola palustris)*

Petty Whin *(Genista anglica)*
Western Gorse *(Ulex gallii)*
Bitter Vetch *(Lathyrus montanus)*

Cloudberry *(Robus chamaemorus)*
Marsh Cinquefoil *(Potentilla palustris)*
Tormentil *(P. erecta)*

Common Sundew *(Drosera rotundifolia)*

Marsh Pennywort *(Hydrocotyle vulgare)*

Sheep's Sorrel *(Rumex acetosella)*

Bog Rosemary *(Andromeda polifolia)*
Bearberry *(Arctostaphylos uva-ursi)*
Ling *(Calluna vulgaris)*
Crossleaved Heath *(Erica tetralix)*
Bell Heather *(E. cinerea)*
Cowberry *(Vaccinium vitis-idaea)*
Bilberry *(V. myrtillus)*
V. × *intermedium*
Cranberry *(V. oxycoccos)*

Crowberry *(Empetrum nigrum)*

Bog Pimpernel *(Anagallis tenella)*

Bogbean *(Menyanthes trifoliata)*

Lousewort *(Pedicularis sylvatica)*

Red Rattle *(P. palustris)*

Pale Butterwort *(Pingicula vulgaris)*

Heath Bedstraw *(Galium saxatile)*
Fen Bedstraw *(G. uliginosum)*

Marsh Valerian *(Valeriana dioica)*

Marsh Thistle *(Cirsium palustre)*
Marsh Dandelion variety *(Taraxacum palustre* agg.*)*

Bog Pondweed *(Potamogeton polygonifolius)*

Bog Asphodel *(Narthecium ossifragum)*

Heath Rush *(Juncus squarrosus)*
Soft Rush *(J. effusus)*
Compact Rush *(J. conglomeratus)*
Bulbous Rush *(J. bulbosus)*
Heath Woodrush *(Luzula multiflora)*

Lesser Twayblade *(Listera cordata)*
Heath Spotted Orchid *(Dactylorhiza maculata* subsp. *ericetorum)*

Common Cotton Grass *(Eriophorum angustifolium)*
Harestail *(E. vaginatum)*
Deer-grass *(Scirpus caespitosus)*
Few-flowered Spike-rush *(Eleocharis pauciflora)*
Common Yellow Sedge *(Carex demissa)*
Smooth Sedge *(C. laevigata)*
Tawny Sedge *(C. hostiana)*
Moor Sedge *(C. binervis)*
Carnation Sedge *(C. panicea)*
Glaucous Sedge *(C. flacca)*
Common Sedge *(C. nigra)*
Star Sedge *(C. echinata)*

White Sedge *(C. curta)*
Flea Sedge *(C. pulicaris)*
Separate-headed Sedge *(C. dioica)*

Purple Moor-grass *(Molinia caerulea)*
Waved Hair-grass *(Deschampsia flexuosa)*
Brown Bent *(Agrostis canina)*
Fine Bent *(A. tenuis)*
Mat-Grass *(Nardus stricta)*

OAK-BIRCH WOODLAND ON THE GRITS AND SHALES

Woodlands in which Sessile Oak *(Quercus petraea)* is dominant are regarded as the climax vegetation on the siliceous soils of the grits and shales. Silver Birch *(Betula pendula)* is usually an abundant associate of the oak and often extends above the altitudinal limit of the latter, while the Hairy Birch *(Betula pubescens)* is also widespread, though not quite so common as its relative and is normally at its best on somewhat wetter ground.

Well-developed semi-natural woodlands of this type are not

extensive and, such as they are, occur in the Upper Derwent Valley, in the Ladybower region, and around the Burbage Brook near Grindleford. On the western side of The Peak there are small areas of such woodland in the upper Dane Valley, and the Oak-Birch association also occurs on both slopes of Longdendale. In addition to the Birches, which are the chief associates of the Oak; Alder *(Alnus glutinosa)* and small quantities of Wych Elm *(Ulmus glabra)* and Ash *(Fraxinus excelsior)* are also found. These species tend to disappear from the community at higher altitudes, however, and the only tree which retains its frequency is the Mountain Ash *(Sorbus aucuparia).*

The herbaceous flora of the Oak-Birch woodlands is naturally of a calcifuge character, but there is a considerable range of species varying with the ecological conditions operative within each woodland. The most important single factor influencing the flora is undoubtedly the water content of the soil which varies considerably from wet marshy places beside the springs and mountain streams to relatively dry rocky places where only a thin layer of acidic humus lies between the outcropping rocks. Quite naturally there are many plants here which find a place in the flora both of the siliceous grassland and the moorlands, but in addition we have a variety of species which are true associates of the woods. Societies of Bluebell *(Endymion non-scriptus)* and Dog's Mercury *(Mercurialis perennis)* are common; Wood Anemone *(Anemone nemorosa)*, Yellow Archangel *(Galeobdolon luteum)*, Water Avens *(Geum rivale)* and the Wild Arum, 'Lords-and-Ladies' *(Arum maculatum)* are also frequent, more especially where conditions are damp and there is a good layer of mild humus. In drier places the ground layer is often dominated by the Bracken to the exclusion of many other species, although there are often complementary societies of plants like the Bluebell in company with the fern. Such associations are successful because the early-flowering Bluebell more or less completes its cycle of growth before the unfolding fronds of Bracken cast too much shade. Certain grasses are widely distributed in Oak-Birch woodlands, and of these Waved Hair-Grass *(Deschampsia flexuosa)* and Soft Wood Grass *(Holcus mollis)* are found in the drier habitats, while Hairy Brome *(Bromus ramosus)*, Giant Fescue *(Festuca gigantea)* and Wood Millet Grass *(Milium effusum)* occur in varying quantities in the damper ones.

The chief constituents of the herbaceous flora of the Oak-Birch woods of the Peak District are given in the following list:

Bracken *(Pteridium aquilinum)*
Hard Fern *(Blechnum spicant)*
Lady Fern *(Athyrium filix-femina)*
Male Fern *(Dryopteris filix-mas)*
Common Buckler Fern *(D. dilatata)*
Beech Fern *(Thelypteris phegopteris)*
Oak Fern *(T. dryopteris)*

Wood Anemone *(Anemone nemorosa)*
Wood Goldilocks *(Ranunculus auricomus)*
Lesser Celandine *(R. ficaria)*

Climbing Corydalis *(Corydalis claviculata)*

Wavy Bittercress *(Cardamine flexuosa)*

Red Gampion *(Silene dioica)*
Greater Stitchwort *(Stellaria holostea)*
Three-veined Sandwort *(Moehringia trinervia)*

Herb Robert *(Geranium robertianum)*

Wood-sorrel *(Oxalis acetosella)*

Bush Vetch *(Vicia sepium)*

Meadowsweet *(Filipendula ulmaria)*

Wild Strawberry *(Fragaria vesca)*
Herb Bennet *(Geum urbanum)*
Water Avens *(G. rivale)*

Golden Saxifrage *(Chrysosplenium oppositifolium)*

Broad-leaved Willow-herb *(Epilobium montanum)*
Common Enchanter's Nightshade *(Circaea lutetiana)*

Sanicle *(Sanicula europaea)*
Cow Parsley *(Anthriscus sylvestris)*
Sweet Cicely *(Myrrhis odorata)*
Hedge-parsley *(Torilis japonica)*
Pignut *(Conopodium majus)*
Hogweed *(Heracleum sphondylium)*

Dog's Mercury *(Mercurialis perennis)*

Sheep's Sorrel *(Rumex acetosella)*

Primrose *(Primula vulgaris)*
Yellow Pimpernel *(Lysimachia nemorum)*

Water Forget-me-not *(Myosotis scorpioides)*
Wood Forget-me-not *(M. sylvatica)*

Foxglove *(Digitalis purpurea)*
Wood Speedwell *(Veronica montana)*
Birdseye Speedwell *(V. chamaedrys)*

Self-heal *(Prunella vulgaris)*
Betony *(Betonica officinalis)*
Hedge Woundwort *(B. sylvatica)*

Yellow Archangel *(Galeobdolon luteum)*
Common Hemp-nettle *(Galeopsis tetrahit)*
Ground Ivy *(Glechoma hederacea)*
Bugle *(Ajuga reptans)*

Woodruff *(Galium odoratum)*
Goosegrass *(G. aparine)*

Common Valerian *(Valeriana officinalis)*

Golden-rod *(Solidago virgaurea)*
Marsh Thistle *(Cirsium palustre)*
Nipplewort *(Lapsana communis)*
Wall Lettuce *(Mycelis muralis)*
Marsh Hawksbeard *(Crepis paludosa)*

Bluebell *(Endymion non-scriptus)*

Ramsons *(Allium ursinum)*

Soft Rush *(Juncus effusus)*
Great Woodrush *(Luzula sylvatica)*
Hairy Woodrush *(L. pilosa)*

Lords-and-Ladies *(Arum maculatum)*

Smooth Sedge *(Carex laevigata)*
Wood Sedge *(C. sylvatica)*
Glaucous Sedge *(C. flacca)*

Purple Moor-grass *(Molinia caerulea)*
Giant Fescue *(Festuca gigantea)*
Rough Meadow-grass *(Poa trivialis)*
Cocksfoot *(Dactylis glomerata)*
Wood Melick *(Melica uniflora)*
Hairy Brome *(Bromus ramosus)*
Creeping Soft-grass *(Holcus mollis)*
Tufted Hair-grass *(Deschampsia caespitosa)*
Waved Hair-Grass *(D. flexuosa)*
Fine Bent *(Agrostis tenuis)*
Wood Millet *(Milium effusum)*

THE CARBONIFEROUS LIMESTONE

As we have already seen the scenery and soils of the limestone districts of The Peak are completely different from those found on the gritstone and shales, and it is not surprising that there are corresponding differences in the flora. The deep dales, which are the most striking feature of this region, are least affected by man's activities and quite naturally it is here that we find the greatest wealth and variety of species. Many more plants are at home on the relatively alkaline soils of the dale slopes than on the acid soils associated with the siliceous rocks, and

1. Dovedale

2. Combs Moss, a gritstone moor, showing a typical rock-stream gully or 'clough'

3. Kinderscout, showing the flat, peaty, heather-covered surface at 2,000 ft. The edge of the plateau is notched by a torrent called the Kinder Downfall

4. *Above*. Deepdale, near Buxton, showing the influence of jointing in the limestone in producing cliffs or 'scars'. *Below*. Mam Tor, the 'Shivering Mountain', near Castleton. The hill is formed of alternating sandstones and shales, the huge quarry-like exposure of these beds being due to natural slipping over underlying Edale Shales

a

b

5. LIMESTONE PLANTS
a. Herb Paris (*Paris quadrifolia*). *b*. Horseshoe Vetch (*Hippocrepis comosa*). *c*. Yellow Star of Bethlehem (*Gagea lutea*). *d*. Common Helleborine (*Epipactis helleborine*)

c

d

. GRITSTONE PLANTS

Hybrid Bilberry (*Vaccinium × intermedium*)

Cowberry (*Vaccinium vitis-idaea*)

Bog Asphodel (*Narthecium ossifragum*)

Heather (*Calluna vulgaris*)

7. FERNS
Above. Oak Fern (*Thelypteris dryopteris*). *Below*. Limestone Fern (*Thelypteris robertiana*)

a

b

8. LICHEN, MOSS AND LIVERWORTS
a. Lichen on gritstone (*Parmelia saxatilis*). *b*. Liverwort (*Conocephalum conicum*). *c*. Moss on limestone (*Fissidens adianthoides*). *d*. Liverwort (*Lunularia cruciata*)

c

d

the three main types of vegetation on the limestone, climax ash woodland, calcareous scrub and calcareous grassland, possess a wonderfully varied assembly of flowering plants.

The normal ecological succession in such a series of communities is one in which the scrub species, dominated by Hawthorn *(Crataegus monogyna)* and Hazel *(Corylus avellana)*, gradually invade the grassland, followed by the establishment of ash saplings within the scrub leading to the ultimate development of climax ash woodland. Were it possible to remove man's influence entirely, this normal succession would re-establish itself, but at present time the grazing of sheep and cattle on the dale slopes tends to maintain a state of relative equilibrium between the vegetation units. Dr C. E. Moss expressed the opinion that most of the limestone habitats were examples of retrogressive communities in which the ash woodlands were still degenerating into scrub, and the scrub into calcareous grassland. This is undoubtedly true in many instances for there are numerous examples of relict woodland species in open grassland in several of the dales, but there are some cases where the normal succession seems to be taking place once again. The southern slopes of Miller's Dale, opposite Raven's Tor, provide one of the best instances of this, for, although grazing still takes place on the slopes, a marked extension of the scrub has been observed during the past thirty-five years.

ASHWOODS

The success of the Ash *(Fraxinus excelsior)* on the limestone is largely due to its ability to thrive on very shallow soils where there is a high lime content. Furthermore, its tolerance of varying conditions of moisture provides another factor which helps it to flourish in these habitats. At the present time the semi-natural ashwoods of The Peak extend to an altitude of approximately 1,000 feet. Above this the woodland degenerates into scrub and in some places a certain amount of Birch *(Betula pendula)* takes the place of the Ash. The richness of the flora of the ash woodland community is due not only to the favourable nature of the soil but also to the abundance of light which the leafy canopy of the Ash permits to penetrate to the woodland floor. Several layers of vegetation occur within this type of community. The uppermost consists of the climax Ash along with arboreal associates like Wych Elm *(Ulmus glabra)* and Sycamore *(Acer pseudo-platanus)*, which is now abundantly naturalised. Beneath the trees there is usually a well-developed layer of shrubs in which Hawthorn *(Crataegus monogyna)* and Hazel *(Corylus avellana)* are the most frequent species. The third layer consists of a rich assemblage of herbaceous plants with extensive societies of Dog's Mercury *(Mercurialis perennis)* and Wood Anemone *(Anemone nemorosa)*, and finally we have the ground layer of bryophytes and lichens which covers the soil and outcropping rocks. Hurt's Wood in Dovedale is probably the best example of this type of semi-natural ash woodland, while Ravensdale possesses another good stretch on its eastern slopes. Monk's Dale, Monsal Dale and Lathkill Dale also have small areas, though these may not be quite so natural. The content of the species within the strata of such woodlands shows considerable variation from one wood to another, though in every case the flora is remarkably rich as will be seen from the list of plants typical of the habitat which is given below.

In addition to the Ash, which is dominant, most woodlands have some or all of the following tree associates:

Sycamore *(Acer pseudo-platanus)*
Maple *(Acer campestre)*
Rowan *(Sorbus aucuparia)*
Sorbus rupicola
Wych Elm *(Ulmus glabra)*
Silver Birch *(Betula pendula)*
Alder *(Alnus glutinosa)*
Aspen *(Populus tremula)*

Similarly quite a wide range of undershrubs can be found, though they vary greatly in frequency. Hawthorn and Hazel are often co-dominant, as already mentioned, but in addition some of the following usually occur:

Common Buckthorn *(Rhamnus cathartica)*
Wild Roses *(Rosa* spp.*)*
Brambles *(Rubus* spp.*)*
Blackthorn *(Prunus spinosa)*
Bird Cherry *(P. padus)*
Spurge-laurel *(Daphne laureola)*
Mezereon *(D. mezereum)* very rarely
Dogwood *(Thelycrania sanguinea)*
Privet *(Ligustrum vulgare)*
Elder *(Sambucus nigra)*
Guelder Rose *(Virburnum opulus)*

The herbaceous layer naturally contains the greatest number of species, and of these mention should first be made of those which produce local societies within the woodland:

Wood Anemone *(Anemone nemorosa)*
Lesser Celandine *(Ranunculus ficaria)*

Red Campion *(Silene dioica)*

Water Avens *(Geum rivale)*

Dog's Mercury *(Mercurialis perennis)*

Wood Forget-me-not *(Myosotis sylvatica)*

Ground Ivy *(Glechoma hederacea)*

Moschatel *(Adoxa moschatellina)*

Ramsons *(Allium ursinum)*

The following species also occur though some of them are very local in their distribution, and in one or two cases are confined to a single woodland:

Bracken *(Pteridium aquilium)*
Hart's-tongue *(Phyllitis scolopendrium)*
Common Spleenwort *(Asplenium trichomanes)*
Wall-rue *(A. ruta-muraria)*

Bladder Fern *(Cystopteris fragilis)*
Male Fern *(Dryopteris filix-mas)*
Common Buckler Fern *(D. dilatata)*
Hard Shield Fern *(Polystichum aculeatum)*

Wood Goldilocks *(Ranunculus auricomus)*
Columbine *(Aquilegia vulgaris)*

Wavy Bittercress *(Cardamine flexuosa)*
Hairy Bittercress *(C. hirsuta)*
Garlic Mustard *(Alliaria petiolata)*

Sweet Violet *(Viola odorata)*
Hairy Violet *(V. hirta)*
Common Dog Violet *(V. riviniana)*
Wood Dog Violet *(V. reichenbachiana)*

Common St John's Wort *(Hypericum perforatum)*
Hairy St John's Wort *(H. hirsutum)*

Greater Stitchwort *(Stellaria holostea)*
Lesser Stitchwort *(S. graminea)*
Three-veined Sandwort *(Moehringia trinervia)*

Meadow Cranesbill *(Geranium pratense)*
Bloody Cranesbill *(G. sanguineum)*
Dovesfoot Cranesbill *(G. molle)*

Shining Cranesbill *(G. lucidum)*
Herb Robert *(G. robertianum)*

Wood Sorrel *(Oxalis acetosella)*

Bush Vetch *(Vicia sepium)*

Meadowsweet *(Filipendula ulmaria)*
Wild Strawberry *(Fragaria vesca)*
Herb Bennet *(Geum urbanum)*

Golden Saxifrage *(Chrysosplenium oppositifolium)*
Alternate Golden Saxifrage *(C. alternifolium)*

Broad-leaved Willow-herb *(Epilobium montanum)*

Sanicle *(Sanicula europaea)*
Rough Chervil *(Chaerophyllum temulum)*
Cow Parsley *(Anthriscus sylvestris)*
Hedge Parsley *(Torilis japonica)*
Burnet-Saxifrage *(Pimpinella saxifraga)*
Greater Burnet-Saxifrage *(P. major)*
Hogweed *(Heracleum sphondylium)*

Primula × variabilis
Primrose *(Primula vulgaris)*
Yellow Pimpernel *(Lysimachia nemorum)*

Birdseye Speedwell *(Veronica chamaedrys)*

Toothwort *(Lathraea squamaria)*

Marjoram *(Origanum vulgare)*
Wild Thyme *(Thymus drucei)*
Self-heal *(Prunella vulgaris)*
Yellow Archangel *(Galeobdolon luteum)*
Wood-sage *(Teucrium scorodonia)*
Bugle *(Ajuga reptans)*

Giant Bellflower *(Campanula latifolia)*
Nettle-leaved Bellflower *(C. trachelium)*
Harebell *(C. rotundifolia)*

Woodruff *(Galium odoratum)*
Crosswort *(Cruciata chersonensis)*
Lady's Bedstraw *(G. verum)*
Slender Bedstraw *(G. pumilum)*

Common Valerian *(Valeriana officinalis)*

Ploughman's Spikenard *(Inula conyza)*
Golden-rod *(Solidago virgaurea)*
Welted Thistle *(Carduus acanthoides)*

Melancholy Thistle *(Cirsium heterophyllum)*
Greater Knapweed *(Centaurea scabiosa)*
Nipplewort *(Lapsana communis)*
Wall Lettuce *(Mycelis muralis)*

Hawkweed *(Hieracium lachenalii)*

Lily of the Valley *(Convallaria majalis)*
Yellow Star of Bethlehem *(Gagea lutea)*

Hairy Woodrush *(Luzula pilosa)*

Common Helleborine *(Epipactis helleborine)*
Common Twayblade *(Listera ovata)*
Early Purple Orchid *(Orchis mascula)*

Lords-and-Ladies *(Arum maculatum)*

Wood Sedge *(Carex sylvatica)*
Fingered Sedge *(C. digitata)*

Giant Fescue *(Festuca gigantea)*
Sheep's Fescue *(F. ovina)*
Wood Meadow-grass *(Poa nemoralis)*
Rough Meadow-grass *(P. trivialis)*
Cocksfoot *(Dactylis glomerata)*
Wood Melick *(Melica uniflora)*
Mountain Melick *(M. nutans)*
Slender False-brome *(Brachypodium sylvaticum)*
Bearded Couch *(Agropyron caninum)*
False Oat *(Arrhenatherum elatius)*
Yorkshire Fog *(Holcus lanatus)*
Tufted Hair-grass *(Deschampsia aespitosa)*

CALCAREOUS SCRUB

It will be evident from what has already been written that quite irrespective of whether the scrub has developed from calcareous grassland in the course of a normal ecological succession or whether it represents a retrogressive community arising from the degeneration of ash woodland, a considerable number of species from the neighbouring communities will find a place in the flora of the scrub. This is the case in nearly all the examples of scrub vegetation in The Peak. Dovedale, Ravensdale, Monk's Dale and Miller's Dale have characteristic examples, but it is probably in Ravensdale where the gradual transition from one community to another can best be observed. The two chief elements of the community are Hawthorn and Hazel, which are, of course, common in the scrub layer of the woodland. In some instances the Hawthorn is dominant, while in others the two species are co-dominant. As in the scrub layer of the woods there are many shrubby associates growing with Hawthorn and Hazel, and it is interesting to note a large number of Ash saplings where the scrub zone is well established, indicating at least the commencement of the normal ecological succession. The following shrubs are present in many of the examples of calcareous scrub mentioned above.

Common Buckthorn *(Rhamnus cathartica)*

Brambles *(Rubus fruticosus* agg.*)*
Field Rose *(Rosa arvensis)*
Burnet Rose *(R. pimpinellifolia)*
Dog Rose *(R. coriifolia)*
Bird Cherry *(Prunus padus)*
Blackthorn *(P. spinosa)*

Dogwood *(Thelycrania sanguinea)*

Crack Willow *(Salix fragilis)*
Pussy Willow *(S. caprea)*

Privet *(Ligustrum vulgare)*

Elder *(Sambucus nigra)*
Guelder Rose *(Viburnum opulus)*
Honeysuckle *(Lonicera periclymenum)*

The herbaceous plants of the scrub are probably just as numerous as those of the woodland and a comparison of the species lists will show that many of the plants are common to each community. There is a corresponding similarity between the herbaceous flora of the scrub and that of calcareous grassland, and it is doubtful whether any species can be considered exclusive to the scrub. Nevertheless the flora of the zone is a rich one as shown by the following:

Globe Flower *(Trollius europaeus)*
Wood Anemone *(Anemone nemorosa)*
Wood Goldilocks *(Ranunculus auricomus)*
Lesser Celandine *(R. ficaria)*
Lesser Meadow-rue *(Thalictrum minus)*

Hairy Violet *(Viola hirta)*
Common Dog Violet *(V. riviniana)*

Common St John's Wort *(Hypericum perforatum)*
Hairy St John's Wort *(H. hirsutum)*

Red Campion *(Silene dioica)*

Bloody Cranesbill *(Geranium sanguineum)*
Herb Robert *(G. robertianum)*

Wood Sorrel *(Oxalis acetosella)*

Wild Strawberry *(Fragaria vesca)*
Agrimony *(Agrimonia eupatoria)*
Common Lady's Mantle *(Alchemilla glabra)*

Salad Burnet *(Poterium sanguisorba)*

Mossy Saxifrage *(Saxifraga hypnoides)*

Broad-leaved Willow-herb *(Epilobium montanum)*

Common Enchanter's Nightshade *(Circaea lutetiana)*

Sanicle *(Sanicula europaea)*
Burnet-Saxifrage *(Pimpinella saxifraga)*
Greater Burnet-Saxifrage *(P. major)*
Hogweed *(Heracleum sphondylium)*

Dog's Mercury *(Mercurialis perennis)*
Greek Valerian *(Polemonium caeruleum)*

Birdseye Speedwell *(Veronica chamaedrys)*

Marjoram *(Origanum vulgare)*
Wood-sage *(Teucrium scorodonia)*

Giant Bellflower *(Campanula latifolia)*
Nettle-leaved Bellflower *(C. trachelium)*
Harebell *(C. rotundifolia)*

Crosswort *(Cruciata chersonensis)*
Lady's Bedstraw *(Galium verum)*
Slender Bedstraw *(G. pumilum)*

Small Scabious *(Scabiosa columbaria)*

Musk Thistle *(Carduus nutans)*
Welted Thistle *(C. acanthoides)*
Spear Thistle *(Cirsium vulgare)*
Hardhead *(Centaurea nigra)*

Dandelion *(Taraxacum officinale)*

Lily of the Valley *(Convallaria majalis)*
Bluebell *(Endymion non-scriptus)*

Herb Paris *(Paris quadrifolia)*

Hairy Woodrush *(Luzula pilosa)*
Good Friday Grass *(L. campestris)*

Common Helleborine *(Epipactis helleborine)*
Common Twayblade *(Listera ovata)*

Early Purple Orchid *(Orchis mascula)*

Wood Sedge *(Carex sylvatica)*
Glaucous Sedge *(C. flacca)*
Fingered Sedge *(C. digitata)*

Giant Fescue *(Festuca gigantea)*
Sheep's Fescue *(F. ovina)*
Wood Melick *(Melica uniflora)*
Mountain Melick *(Melica nutans)*
Hairy Brome *(Bromus ramosus)*
Tufted Hair-grass *(Deschampsia cespitosa)*

CALCAREOUS GRASSLAND

The grassland community on the Mountain Limestone is much more extensive than either the ash woodland or the scrub and is found in all the dales. In a number of these there is a certain amount of grazing which has a considerable influence on the vegetation, but there are places where this does not take place. Parts of the upper eastern slopes of Monk's Dale are relatively free from such interference, while the northern slopes of Miller's Dale from the village to the vicinity of Ravensdale Tor are entirely ungrazed and, apart from occasional burning of the plant cover, provide as natural a stretch of calcareous grassland as can be found. Plant life here is typically rich in species and the flora of such grasslands possesses greater variety than that of any other single community within the Peak District, for although it includes many species which are constituents of the calcareous habitats already described, there are others which are exclusive to grassland. The dominant plant is Sheep's Fescue Grass *(Festuca ovina)*, which owes its success to its adaptation to dry conditions. Other grasses like *Dactylis glomerata, Briza media, Anthoxanthum odoratum, Helictotrichon pratense* and *Koeleria cristata* are common associates of the *Festuca*, while the less widely distributed species include *Arrhenatherum elatius, Helictotrichon pubescens, Sieglingia decumbens, Cynosurus cristatus*, and *Holcus lanatus*. In areas where there has been much leaching *Festuca ovina* often remains dominant, but such calcifuge species as *Deschampsia flexuosa* and *Agrostis tenuis* form part of the association. A very wide range of species grows in company with the above grasses which form the turf, and an especially rich flora is present in those areas where the grazing is absent. The following plants are present in the grassland community in many of the dales, though the list is by no means complete:

Globe Flower *(Trollius europaeus)*
Meadow Buttercup *(Ranunculus acris)*
Creeping Buttercup *(R. repens)*
Bulbous Buttercup *(R. bulbosus)*
Lesser Meadow-rue *(Thalictrum minus)*

Hutchinsia *(Hornungia petraea)*
Twisted Whitlow-grass *(Draba incana)*
Wall Whitlow-grass *(D. muralis)*

Hairy Bitter-cress (*Cardamine hirsuta)*
Hairy Rock-cress (*Arabis hirsuta)*
Thale Cress (*Arabidopsis thaliana)*

Hairy Violet (*Viola hirta)*
Common Dog Violet (*V. riviniana)*
Mountain Pansy (*V. lutea)*

Common Milkwort (*Polygala vulgaris)*

Elegant St John's Wort (*Hypericum pulchrum)*
Hairy St John's Wort (*H. hirsutum)*
Common Rock-rose (*Helianthemum chamaecistus)*

Nottingham Catchfly (*Silene nutans)*
Maiden Pink (*Dianthus deltoides)*
Thyme-leaved Sandwort (*Arenaria serpyllifolia)*

Fairy Flax (*Linum catharticum)*

Bloody Cranesbill (*Geranium sanguineum)*
Dovesfoot Cranesbill (*G. molle)*
Shining Cranesbill (*G. lucidum)*
Herb Robert (*G. robertianum)*

Black Medick (*Medicago lupulina)*
Red Clover (*Trifolium pratense)*
Zigzag Clover (*T. medium)*
White Clover (*T. repens)*
Kidney Vetch (*Anthyllis vulneraria)*

Common Birdsfoot Trefoil (*Lotus corniculatus)*
Horseshoe Vetch (*Hippocrepis comosa)*
Bush Vetch (*Vicia sepium)*

Dropwort (*Filipendula vulgaris)*
Barren Strawberry (*Potentilla sterilis)*
Spring Cinquefoil (*P. tabernaemontani)*
Tormentil (*P. erecta)*
Parsley Piert (*Aphanes arvensis)*
Salad Burnet (*Poterium sanguisorba)*

Grass of Parnassus (*Parnassia palustris)*

Pignut (*Conopodium majus)*
Burnet-Saxifrage (*Pimpinella saxifraga)*
Greater Burnet-Saxifrage (*P. major)*

Cowslip (*Primula veris)*

Felwort (*Gentianella amarella)*

Greek Valerian (*Polemonium caeruleum)*

Common Forget-me-not (*Mysotis arvensis)*
Early Forget-me-not (*M. ramosissima)*

Birdseye Speedwell (*Veronica chamaedrys)*

Thyme Speedwell *(V. serpyllifolia)*
Yellow Rattle *(Rhinanthus minor* agg.*)*
Eyebright *(Euphrasia nemorosa* and *E. anglica)*
Red Bartsia *(Odontites verna)*

Marjoram *(Origanum vulgare)*
Wild Thyme *(Thymus drucei)*
Self-heal *(Prunella vulgaris)*
Wood-sage *(Teucrium scorodonia)*

Hoary Plantain *(Plantago media)*
Ribwort Plantain *(P. lanceolata)*

Clustered Bellflower *(Campanula glomerata)*
Harebell *(C. rotundifolia)*

Crosswort *(Cruciata chersonensis)*
Lady's Bedstraw *(Galium verum)*
Slender Bedstraw *(G. pumilum)*

Cornsalad *(Valerianella locusta)*
Keeled Cornsalad *(V. carinata)*

Field Scabious *(Knautia arvensis)*
Small Scabious *(Scabiosa columbaria)*
Devilsbit Scabious *(Succisa pratensis)*

Mountain Everlasting *(Antennaria dioica)*
Golden-rod *(Solidago virgaurea)*
Yarrow *(Achillea millefolium)*
Carline Thistle *(Carlina vulgaris)*
Greater Knapweed *(Centaurea scabiosa)*
Hardhead *(C. nigra)*
Sawwort *(Serratula tinctoria)*
Autumn Hawkbit *(Leontodon autumnalis)*
Greater Hawkbit *(L. hispidus)*
Hawkweed Ox-Tongue *(Picris hieracioides)*
Hawkweeds
Hieracium lachenalii
H. britannicum
Mouse-ear Hawkweed *(H. pilosella)*
Dandelion *(Taraxacum officinale)*

Lily of the Valley *(Convallaria majalis)*

Common Helleborine *(Epipactis helleborine)*
Dark Red Helleborine *(E. atrorubens)*
Frog Orchid *(Coeloglossum viride)*
Scented Orchid *(Gymnadenia conopsea)*
Dwarf Burnt Orchid *(Orchis ustulata)*
Early Purple Orchid *(Orchis mascula)*
Common Spotted Orchid *(Dactylorhiza fuchsii)*
Pyramid Orchid *(Anacamptis pyramidalis)*

Carnation Sedge *(Carex panicea)*

Glaucous Sedge *(C. flacca)*
Spring Sedge *(C. caryophyllea)*
Small Fingered Sedge *(C. ornithopoda)*

Smooth Meadow-grass *(Poa pratensis)*
Mountain Melick *(Melica nutans)*
Slender False-brome *(Brachypodium sylvaticum)*
Tufted Hair-grass *(Deschampsia caespitosa)*
Meadow Foxtail *(Alopecurus pratensis)*

Within the grassland community there are several local societies of considerable interest. Broken ground resulting from old lead-mine working or prospecting occurs in many places on the limestone, and on the detritus of which these sites are formed there are one or two plants worthy of special note. Of these, Vernal Sandwort *(Minuartia verna)* is probably the most striking. It is a low-growing species with a cushiony habit and a wealth of conspicuous white flowers, and its association with the lead-mine spoil-heaps has given it the local name of Leadwort, for it is seldom seen elsewhere in the Peak District. Another species intimately associated with the same type of locality is a form of Alpine Pennycress *(Thlaspi virens)*, which was known only in Derbyshire until quite recently. In the Alport district similar ground is colonised by the beautiful Black Mullein *(Verbascum nigrum)*, though probably no direct connection exists between this species and the habitat conditions since in other parts of Britain it is a typical plant of calcareous ground where no lead-mining has taken place. On the southern side of Miller's Dale, above the stretch of the Wye Valley known locally as Water-cum-Jolly, ground associated with old lead workings supports a number of species which attract botanists from many parts of the country. The most notable is undoubtedly the unusual Birds'-foot Sedge *(Carex ornithopoda)*, for this is the 'locus classicus' of this species in Britain, it being first discovered here in 1873 and described for the British flora from material gathered at this spot. *Carex ornithopoda* still occurs in its original habitat but in addition is now known in nearby Ravensdale on sites unconnected with the lead-mine activities. It still remains one of our rarest Sedges for there are only three other vice-counties in which it has been found. Associated with the *Carex* in the Miller's Dale site one can find another rarity, the Twisted Podded Whitlow-Grass *(Draba incana)*, and an abundance of the Lesser Meadow-Rue *(Thalictrum minus)*.

Many of the dales possess a wealth of rocky outcrops which form shelves or ledges with only a thin covering of soil. These often have a little plant association of their own. Such tiny species as the Vernal Whitlow-Grass *(Erophila verna)*, Hutchinsia *(Hornungia petraea)*, Early Forget-me-not *(Myosotis ramosissima)*, Dove's-foot Cranesbill *(Geranium molle)*, Stork's Bill *(Erodium cicutarium)*, Parsley Piert *(Aphanes arvensis)*, and Spring Cinquefoil *(Potentilla tabernaemontani)* are present in some quantity during the spring and early summer but later become shrivelled as the soil dries out. *Hornungia* is of particular interest, for its frequency varies considerably from year to year, and the visiting botanist who chances to see it in an optimum year may take away a completely erroneous picture of its status in the district. The reasons for the periodicity of this annual species are not yet fully understood, but in seasons when conditions are favourable the seeds germinate in the autumn and the seedlings over-winter, but severe weather conditions in the early months of the year may cause many of these to be lost with a consequent scarcity of the plant in the following season and possibly for several subsequent seasons. Careful field observations have shown that autumn germination is not invariably the case, and it would seem that where conditions cause the germination of a high percentage of the seeds to be delayed until the spring the chances of survival of the seedlings are distinctly improved and a good year may ensue.

The vertical faces of the rocky outcrops possess a wealth of fissures and crannies in which small ferns often grow in abundance. In the driest corners where there seems to be little or no soil, the Wall Rue Spleenwort *(Asplenium ruta-muraria)* flourishes, and in somewhat moister spots the Maidenhair Spleenwort *(Asplenium trichomanes)* is common. The rare Green Spleenwort *(Asplenium viride)* also occurs, but it seems that this species requires damper and shadier conditions for it invariably inhabits the deeper crannies and prefers the north-facing slopes of the dales. The curious Scaly Spleenwort, or Rusty-back Fern *(Ceterach officinarum)* is now known to be increasing in several localities – a gratifying change of status, for at the turn of the century it became almost extinct in the Peak District on account of the predatory activities of fern gatherers who sold this and other attractive species on the market-places of several of The Peak towns. Perhaps the most delicate and handsome fern of all is the Brittle Bladder Fern *(Cystopteris fragilis)*, which is widespread on most of the limestone cliffs both large and small. Quite a number of Hawkweeds *(Hieracia)* are found on the outcropping rocks in addition to the ferns, and of these *H. britannicum* is a typical species. Less common Hawkweeds in such habitats are *H. pellucidum* and *H. exotericum*.

Beneath the larger limestone cliffs in any of the dales there are extensive screes formed of fragments of rock of varying size weathered from the exposed masses above. Such areas constitute edaphic deserts. There are no true lithophytes in these localities but certain plants gradually colonise the rocks. Lichens and bryophytes pave the way for higher plants, and of the latter two species are of special note. The Limestone Fern *(Thelypteris robertiana)* can often be seen growing through the stones and frequently the Herb Robert *(Geranium robertianum)*, which turns a lovely red as the season advances, forms a colourful fringe round the edge of the scree. As the process of colonisation of these habitats continues a thin layer of plant debris and blown soil accumulates and other species gradually enter the community. The Stone Bramble *(Rubus saxatilis)* is a characteristic plant, and the delicate Rock Rose *(Helianthemum chamaecistus)*, the Wild Thyme *(Thymus drucei)*, and the Bloody Cranesbill *(Geranium sanguineum)* may become a feature of the association.

LIMESTONE RIVERS AND STREAMS

As one might expect, the plant life of the mountain streams which flow through many of the limestone dales is as interesting as that of other Peakland habitats. Those rivers which are remote from towns and villages, at least in their upper reaches, are quite free from pollution and have a rich flora of submerged and marginal species, though plants with floating leaves, such as Water Lilies, are not common on account of the swiftness of the waters. Such rivers as the Dove, the Lathkill and the Bradford are the most typical examples. Water Buttercups are well represented and *Ranunculus trichophyllus*, which is the earliest to flower, occurs in streams with a somewhat gentler rate of flow. *Ranunculus aquatilis* subsp. *peltatus*, with floating leaves, also occurs, but the most abundant species, especially in the faster-flowing waters, is *Ranunculus aquatilis* subsp. *pseudo-fluitans*, a plant whose long stems and crowded, tassel-like leaves sway gently to and fro as the sparkling streams ripple above them. In the lower reaches of the River Dove below Thorpe Cloud there is a tendency for *Ranunculus fluitans* to replace the previous species. Another plant which is abundant and often completely submerged is the Procumbent Marshwort *(Apium nodiflorum)* while the Pondweeds are represented by the Broad-leaved Pondweed *(Potamogeton natans)*, the Curled-leaved Pondweed *(P. crispus)*, and one of the Small Pondweeds *(P. berchtoldii)*.

The flora of the river margins includes quite a range of reed-swamp plants like the Branched Burr-reed *(Sparganium erectum)*, the Yellow Flag *(Iris pseudacorus)*, the Purple Loosestrife *(Lythrum salicaria)*, and a number of Sedges like *Carex acutiformis*, *C. rostrata*, and *C. acuta*. Associated with these are such moisture-loving species as the Marsh Marigold *(Caltha palustris)*, Skullcap *(Scutellaria galericulata)*, the Water Parsnip *(Berula erecta)*, Water Figwort *(Scrophularia aquatica)*, Square-stalked St John's Wort *(Hypericum tetrapterum)*, Meadowsweet *(Filipendula ulmaria)*, Brooklime *(Veronica beccabunga)*, and Water Speedwell *(V. anagallis-aquatica)* as well as several mints like the Water Mint *(Mentha aquatica)* and Peppermint *(M. ×piperita)*.

REFERENCES

Linton, W. R. *Flora of Derbyshire*. Bemrose (1903)
Moss, C. E. *Vegetation of the Peak District*. Cambridge University Press (1913)
Tansley, A. G. *Britain's Green Mantle*. Allen and Unwin (1949)
Tansley, A. G. *The British Islands and their Vegetation* (2 vols). Cambridge University Press (1949)
Pearsall, W. H. *Mountains and Moorlands*. Collins: New Naturalist (1950)
Balme, O. E. Edaphic and Vegetational Zoning on the Carboniferous Limestone of the Derbyshire Dales. *Jour. Ecology: 41* (1953)
Matthews, J. R. *Origin and Distribution of the British Flora*. Hutchinson (1955)
Grime, J. P. An Ecological Investigation at a junction between two communities in Coombsdale on Derbyshire Limestones. *Jour. Ecology: 51* (1963)
Eyre, S. R. Vegetation of a S. Pennine Upland. *Geography as Human Ecology*. ed. Eyre and Jones. Arnold (1966)
Lloyd, P. S. The ecological significance of fire in limestone grassland communities of the Derbyshire Dales. (1968)
Shimwell, D. Vegetation of the Derbyshire Dales. *Nature Conservancy Report* (1968)
Clapham, A. R. (ed.) *Flora of Derbyshire*. (1969)

7. Flowerless Plants

> Gravestones do not last for ever, nor for very long; in the meantime Nature is . . . touching, softening and tinging them with her mosses.
>
> *W. H. Hudson: Hampshire Days*

MOSSES AND LIVERWORTS

Although the Bryophytes, which include mosses *(Musci)* and liverworts *(Hepaticae)*, are by no means as spectacular as the flowering plants they are none the less an important element in the flora of any area. In much the same way as the higher plants, their distribution and range of species is intimately related to the conditions of rock and soil, and in the Peak District the presence of both siliceous and calcareous rocks results in a much richer bryophyte flora than would otherwise occur.

These lowly plants show marked preferences for particular habitats. Some grow extensively in grassland communities; others prefer the damper and more shady conditions of the woodland floor; a considerable number thrive in moist crannies of the rocks, while there are those which can survive exposure to sun and drying winds on the tops of walls and on open screes. In common with the flowering plants bryophytes are influenced by the varying degrees of alkalinity and acidity of the soil. Some genera, such as *Sphagnum*, can only be found where conditions are decidedly acid, and are therefore widespread in the wet moorland bogs; similarly there are genera which demand an alkaline soil and so thrive on steep limestone slopes and on outcropping rocks. Although a much greater range of species is to be found on the limestone than on the gritstone, the combination of the two floras in The Peak provides a substantial list of bryophytes which includes quite a number of rare species as well as one moss which is unique.

At the beginning of the century over 300 mosses and nearly 100 liverworts or hepatics were known in the Peak District. Since that time several additions have been made to the records of these plants but this increase is largely offset by the fact that a number of species have not been seen for many years. It seems likely, therefore, that the total number remains much the same as it was about sixty years ago. That so many bryophytes do occur is accounted for by the great variety of habitats provided by the different rock formations, by the relatively high rainfall, and by the considerable range of altitude.

The Gritstone Moors

On the gritstones and shales where acid soil conditions prevail both mosses and liverworts are well represented although the number of species is more limited than upon the limestone. In the wetter parts of the high moorlands various species of *Sphagnum* occur which are often associated with the common hair moss, *Polytrichum commune*, while in somewhat drier areas the smaller species *P. piliferum, P. aloides,* and *P. juniperinum* are widespread. During late spring the red star-like male inflorescence of *P. piliferum* is a strikingly characteristic feature of dry banks and moorland footpaths. Typical mountain species such as *P. alpinum* and *P. strictum* have been recorded on Kinderscout in the past but there are no records of them having been seen in recent years. This rarity of mountain bryophytes in the High Peak is also revealed in the relative infrequence of members of the genus *Andreaea*, which are so common on the face of siliceous rocks in regions like the Lake District and North Wales. *Andreaea rothii* and *A. crassinervia* still survive in The Peak but only in very small quantity. Other mountain species which have not been recorded recently include *Blindia acuta* and *Hedwigia ciliata*, and the question arises as to whether the possible disappearance of such plants does not bear some relationship to the increasing atmospheric pollution from neighbbouring industrial areas. In contrast to the apparent disappearance of certain bryophytes from the Peak District at least one species has established itself during the last quarter of a century. *Orthodontium lineare* was first recorded on Kinderscout in 1930 and by 1950 it had become one of our commonest mosses on peaty banks and rotting tree-stumps, not only on Kinder but in other acid habitats throughout the district.

The moss and liverwort flora of the streams which fall from the high moorlands is often luxuriant although the number of species may not be numerous. Certain plants provide a colourful carpet beside the tumbling water, while others clothe the dripping rocks with an equally striking fringe. As examples, the localities red-brown patches of *Marsupella emarginata* colour the wet rocks and overhanging ledges, while occasionally the purple covering of *Nardia compressa* produces an even richer effect.

In addition to the bryophytes already referred to, the following species form part of the moorland flora:

MOSSES

Tetraphis pellucida
Oligotrichum hercynicum
Dicranella crispa
Dicranella rufescens
D. heteromalla
Campylopus pyriformis
C. fragilis
Dicranum scoparium
Rhacomitrium lanuginosum
Leptodontium flexifolium
Aulacomnium palustre
Pohlia nutans
Bryum pallens
Philonotis fontana
P. caespitosa
Drepanocladus fluitans
D. revolvens
Cratoneuron commutaum var. *falcatum*
Hygrohypnum ochraceum
Hyocomium flagellare

LIVERWORTS

Ptilidium ciliare
Lepidozia reptans
Calypogeia trichomanis
C. arguta
C. fissa
Cephalozia bicuspidata
Scapania nemorosa
Mylia anomala
Mylia taylori
Gymnocolea inflata
Barbilophozia floerkei
Lophozia ventricosa
L. incisa
Solenostoma cordifolium
S. sphaerocarpum
S. crenulatum
Plectocolea obovata
Diplophyllum albicans

Siliceous Woodland

The somewhat fragmentary oakwoods in which *Quercus petraea* is dominant often have a luxuriant byrophyte flora, though again the number of species may be restricted. Plants like *Atrichum undulatum* and *Mnium hornum* are widespread on the woodland floor, and there is usually an abundance of *Tetraphis pellucida* at the base of the trees where it is commonly associated with the beautiful hepatic *Lepidozia reptans*. *Dicranum scoparium* is abundant and both *Plagiothecium denticulatum* and *P. undulatum* are common in many places. The long flattened strands of the latter are a particularly conspicuous feature of such acid habitats, and are so characteristic as to render the moss readily recognisable even to the inexpert eye.

Although an abundance of mosses and liverworts can be seen on the ground and at the base of the trees it is surprising that throughout the Peak District tree-trunks and branches exhibit remarkably few of the bryophytes which are usually associated with them. For instance, examples of the genus *Ulota* are seldom met with and very few species of *Orthotrichum* are to be found. This scarcity of arboreal bryophytes is even more true of the lichens, and this may be further evidence of restriction due to the effects of atmospheric pollution.

The following list of species, though not an exhaustive one, gives some idea of the flora of the siliceous woodlands:

MOSSES

Tetraphis pellucida
Atrichum undulatum
Polytrichum formosum
Dicranella hetromalla
Dicranum scoparium
D. majus
Leucobryum glaucum
Fissidens bryoides
F. adianthoides
F. taxifolius
Bryum capillare
Mnium hornum
M. undulatum
M. stellare
M. punctatum
Thamnium alopecurum

Eurhynchium praelongum
E. striatum
E. confertum
Plagiothecium denticulatum
P. undulatum
P. elegans
Pleurozium schreberi
Hylocomium splendens

LIVERWORTS

Lepidozia reptans
Bazzania trilobata
Ptilidium ciliare
Diplophyllum albicans
Calypogeia trichomanis
C. arguta
Cephalozia bicuspidata
Scapania nemorosa
Lophocolea bidentata
L. cuspidata
L. heterophylla
Plagiochila asplenioides
Solenostoma sphaerocarpum
Lophozia ventricosa
Nardia scalaris
Pellia epiphylla
Conocephalum conicum

The Limestone

In common with the higher plants the proportion of calcicole bryophytes is greater than that of calcifuge species, and this fact, coupled with the considerable variety of habitat found on the limestone, has resulted in quite a rich flora especially in the dales.

The bryophyte flora of the semi-natural ash woodland bears a close resemblance to that of the calcareous scrub, and is made up of species which grow on the soil, those growing at the base of trees and shrubs, and those which cover the limestone boulders which are strewn about the woodland floor. *Atrichum undulatum* and *Mnium hornum* are especially common on the soil, while *Hypnum cupressiforme* and *Tetraphis pellucida* often occur in some quantity around the tree bases. The commonest species to be found on the bark of the ash is *Dicranoweisia cirrata*, occasionally in association with the rare *Dicranum montanum*, while on the north face of such trees there may be a considerable growth of the leafy liverwort *Lophocolea cuspidata*. Stones lying beneath the trees support the richest bryophyte flora in the woodland. Boulders are sometimes almost hidden by the lovely moss *Thuidium tamariscinum*, and the characteristic limestone species *Ctenidium molluscum* is even more common. *Brachythecium populeum*, *Eurhynchium praelongum*, and *Amblystegium serpens* are frequent and there is often an abundance of the hepatic *Plagiochila aspleniodes*. Growing on the earth among the stones and outcropping rocks, particularly under moist and shady conditions, a miniature carpet of 'thallose' liverworts may frequently develop. Of these, *Conocephalum conicum* is usually the most conspicuous, but it is often in company with *Marchantia polymorpha*, *Lunularia cruciata* and *Pellia fabbroniana*. The following species have also been found, though not all of them in the habitats just described:

MOSSES

Dicranella heteromalla
Fissidens cristatus
F. bryoides
F. adianthoides
Tortella tortuosa
Encalypta streptocarpa
Orthotrichum anomalum, var. *saxatile*
Aulacomnium androgynum
Bryum capillare
Rhodobryum roseum
Mnium undulatum
M. rostratum
Camptothecium sericeum
Brachythecium rutabulum
B. populeum
B. velutinum
Cirriphyllum crassinervium
Pseudoscleropodium purum
Isothecium myosuroides
Acrocladium cuspidatum
Hylocomium splendens
Rhytidiadelphus squarrosus
R. triquetrus
Rhytidium rugosum
Thamnium alopecurum
Drepanocladus uncinatus
Tortula subulata

LIVERWORTS

Lophocolea bidentata

Porella platyphylla
Lejeunea cavifolia
Cephalozia bicuspidata
Calypogeia trichomanis
Metzgeria pubescens
M. furcata

The limestone crags which contribute so much to the character and beauty of the dales, and the massive cliffs which line ravines like Chee Dale and Ravensdale provide the richest bryophyte habitats in The Peak. The flora varies considerably according to the degree of moisture and shade but it is not surprising that damp conditions like those in Ravensdale or in the Dove Holes (Dovedale) enable the widest range of species to flourish. Even so, on the drier and more exposed rocks there are many species well adapted to the conditions. Cushions of *Grimmia pulvinata* are abundant and *G. apocarpa* is often equally common. *Tortula muralis, T. ruralis, T. intermedia,* and *Tortella tortuosa* occur frequently, and *Trichostomum crispulum* can usually be found. Exposed rock surfaces are abundantly clothed with the golden-green strands of *Camptothecium sericeum,* while in the spring it is not unusual to find the striking extinguisher-like fruits of *Encalypta vulgaris* in crevices. The liverworts on drier surfaces are less abundant but *Porella platyphylla* is characteristic and in some places *Frullania tamarisci* occurs, forming reddish-brown patches on the rock.

A careful examination of the damper cliff-faces, especially in shady ravines, quickly reveals species of mosses and liverworts which have not been observed elsewhere. Not infrequently a dark green film on a damp surface is due to one or other of the species of *Seligeria* which are our smallest acrocarpous mosses. In Ravensdale alone three species may occur almost side by side, *Seligeria acutifolia* var. *longiseta, S. pusilla* and *S. doniana.* Sometimes these are in company with another rarity, *Gymnostomum calcareum,* a species which almost rivals *Seligeria* in smallness. In both Ravensdale and Dovedale this moss has been seen in association with one of the most minute and delicate of the pleurocarpous species, for the tiny strands of *Amblystegiella sprucei* occasionally trail over the *Gymnostomum.* Pure patches of *Amblystegiella* are of rare occurrence though good material can be found on tufa in Monk's Dale.

Both thallose and leafy liverworts are well represented in the damp ravines, and of the former *Conocephalum conicum, Reboulia hemisphaerica, Lunularia cruciata,* and *Pellia fab-*

broniana are the commonest, with a local abundance of *Metzgeria furcata,* and *M. pubescens* in certain places. Quite a number of leafy species occur and of these two members of the genus *Cololejeunea* are perhaps the most notable: *C. rossettiana* and *C. calcarea.* While in Britain as a whole the former is the more rare, in the Peak District the converse is true. These plants are readily overlooked by reason of their small size, but a careful scrutiny of the rock-faces in several of the narrow dales will reveal their presence, in some cases attached to the bare rock and in others trailing over larger bryophytes. In one small ravine not far from Buxton both species can be found. Here in the lower and more shady parts only *Cololejeunea rossettiana* occurs, but higher up under conditions of better light it gives place to *C. calcarea.* The relation between shade conditions and the occurrence of these two species of *Cololejeunea* in this habitat may not be of any ecological significance, though it is a matter which might repay investigation.

One of the limestone ravines in north Derbyshire has claim to bryological importance on account of a species which is unique to that locality. *Thamnium augustifolium* was discovered at the source of an intermittent spring in 1883 by G. A. Holt. Its morphological characters were found to be sufficiently distinct to separate it from its common relative *T. alopecurum* and a new species was recognised. No other site for it has been discovered anywhere in the world, and modern bryologists are uncertain whether the distinct differences of habit and leaf structure are not in reality the result of environmental variations rather than of genetic origin. Future cultural experiments may solve this problem.

A few species of mosses and liverworts form local societies on rocky shelves. The latter often have a thin covering of soil which supports mosses like *Pottia recta, P. bryoides, P. lanceolata, P. starkeana, Phascum cuspidatum, Funaria muehlenbergii,* and occasionally *F. fascicularis. Tortula muralis, T. subulata, Tortella turtuosa* and *Trichostomum crispulum* are frequently present, and *Plagiobryum zeirii,* with its vinous lower leaves, is perhaps the most attractive species in this habitat. During spring and early summer, while the shallow layer of soil is still moist, two species of thallose liverwort of particular interest occur. These are *Riccia sorocarpa,* which has been observed in such habitats in many of the dales, and *Targionia hypophylla,* which is commoner than the published records indicate. This is understandable for both species shrivel and disappear as soon as the sun dries out the soil, and at the height of the summer their presence can no longer be detected.

In some of the damp limestone valleys deposits of tufa are found to be closely connected with the mosses that grow upon them. Certain species appear to have the capacity to precipitate calcium carbonate from the water which is constantly dripping over them, thus forming beds of tufa of varying thickness. This phenomenon is well-developed in Ravensdale, where an attenuated form of *Cratoneuron commutatum* has built up a wall of tufa two feet or more high and several feet long. In the same locality *Eucladium verticillatum* and *Barbula tophacea* show similar tufa-forming characters as does the thallose liverwort *Pellia fabbroniana*. The exact process which causes the deposition of the calcium carbonate is not fully understood, but in the case of the *Pellia* it is evidently brought about by an alkaline excretion which precipitates the calcium carbonate from the bicarbonate solution.

The bryophyte flora of the limestone cliffs and rocky outcrops quite naturally includes many species which have not been commented upon and some of the more characteristic are given in the following list:

MOSSES

Fissodens viridulus
F. cristatus
F. pusillus
Barbula fallax
B. vinealis
B. recurvirostris
Trichostomum brachydontium
T. brachydontium var. *cophocarpum*
Neckera crispa
N. complanata
Anomodon viticulosus
Brachythecium populeum
Eurhynchium praelongum
E. swartzii
E. murale
Rhyncostegiella pallidirostra
R. teesdalei
R. tenella
Isopterygium depressum
Cratoneuron filicinum
Amblystegium serpens
Encalypta streptocarpa

Orthotrichum anomalum var. *saxatile*
O. cupulatum
Mnium stellare
M. rostratum

LIVERWORTS

Frullania tamarisci
F. dilatata
Lejeunea cavifolia
Scapania aequiloba
S. aspera
Pedinophyllum interruptum
Plagiochila asplenioides
Solenostoma triste
Tritomaria quinquidentata
Reboulia hemisphaerica
Riccardia multifida

The Limestone Streams

Clear sparkling streams are a feature of many of the limestone dales and these have their characteristic bryophytes. In some of them there is an abundance of *Fontinalis antipyretica*, a striking moss which is commonly attached to submerged rocks just beneath the water surface. It is somewhat variable in form but normally the leaves are arranged in three well-marked ranks which give the leafy strands a typical triangular (triquetrous) appearance, making it easily recognisable even by the non-bryologist. A remarkably beautiful variety of this species occurs in Lathkill Dale, namely *F. antipyretica* var. *cymbifolia*, in which the leaves are more markedly boat-shaped and spread at right-angles to the stem in contrast to the three tight ranks of the more typical plant, giving the moss a completely different appearance. It has been suggested that var. *cymbifolia* is a still-water form, but this is difficult to reconcile with the fact that normal plants grow nearby in waters which are at least as slow flowing.

The amount of water flowing in the limestone streams is variable from one season to another. Thus the rocks which form the bed are not everywhere continuously submerged, and such conditions are particularly suitable for the growth of certain mosses. In Monk's Dale one can find species like *Cratoneuron commutatum* var. *falcatum*, *Bryum pseudotriquetrum*, *Philonotis fontana* and *P. calcarea* growing on the stones, and in many dales there is an abundance of *Cinclidotus fontinaloides*. *Fissidens crassipes* is one of the less common mosses on boulders in and around the streams, though its frequency varies considerably from year to year in places where it is known to grow. For example, it was extremely abundant at the foot of Ravensdale during the period of drought towards the end of the summer of 1949, but since then it has only occurred in this site in relatively small quantity.

Hepatics of the limestone streams are much less common and only *Chiloscyphus polyanthus* var. *rivularis* is really abundant. In the stream which flows through Ravensdale there is a particularly rich growth on submerged stones.

Calcareous Grassland

Many bryophytes which occur in the woodland and scrub zones also find a place in the flora on the calcareous grassland, and only a few species can be considered to belong exclusively to the grassland. On most of the grassy limestone slopes *Pseudoscleropodium purum*, *Camptothecium lutescens*, *Dicranum scoparium* and *D. bonjeani* are quickly detected. *Hypnum cupressiforme* in its larger forms is widespread, and both *Hylocomium splendens* and *Pleurozium schreberi* are common associates. Of the genus *Thiudium* only *T. tamariscinum* is common in the Peak District though *T. abietinum* has been recorded. *Rhytidiadelphus triquetrus* and *R. squarrosus* are plentiful, while in the somewhat stony parts of the grassland *Rhytidium rugosum* is not uncommon. This handsome moss is somewhat rare in Britain as a whole, but it seems to flourish in sites which represent the later stages in the colonisation of screes. Other moss species which occur fairly commonly includes the following:

Campylium chrysophyllum
Ditrichum flexicaule
Eurhynchium swartzii
Cirriphyllum crassinervium
Fissidens cristatus
Mnium undulatum
Tortula intermedia
Barbula recurvirostris
Weissia controversa
Encalypta streptocarpa
Rhyncostegiella tenella

Isothecium myosuroides
Tortella tortuosa
Trichostomum crispulum
Orthothecium intricatum
Brachythecium rutabulum
B. velutinum
B. glareosum

Hepaticae are not abundant in the limestone grassland of The Peak but the following have been observed:

Plagiochila asplenioides
Scapania aspera
S. aequiloba
Reboulia hemisphaerica
Lophocolea bidentata
Leiocolea muelleri

Ponds and Lakes

One of the most fascinating habitats for bryophytes is the shore mud of a pond or lake during a period of drought. Such places in The Peak were closely studied by the late Professor T. Barker towards the end of the nineteenth century and valuable records were made. This able bryologist lived at Whaley Bridge and was therefore conveniently placed to make frequent observations on the bryophytes associated with the shores of both Combs Lake and Toddbrook Reservoir. The common feature of bare mud is an abundant growth of minute species of ephemeral mosses which develop and mature in a very short time, and Professor Barker's most important discovery was that of *Physcomitrium sphaericum,* one of the rarest of these annuals, both at Combs Lake and Toddbrook Reservoir. His record for the species at Combs Lake was made in 1894 and there is no evidence in bryological literature of it having been seen again until rediscovered by the writer during the drought of 1947. Although this does not necessarily imply that the moss did not occur during the intervening years it is well known that such plants tend to disappear for considerable periods while conditions are adverse. Since these annuals depend upon spore germination for their continued existence, it is of interest to speculate upon the way in which they bridge the years when ecological conditions are unfavourable. Possibly the spores retain their viability for a considerable time whilst submerged and germinate only when the water recedes and a subsequent drought uncovers the shore mud once more, or alternatively the spores may germinate in the water and produce a protonema which persists until a similar recession of the water produces conditions which enable the gametophyte to mature.

In 1947, from mid-July until early November, conditions were ideal for the growth of ephemeral mosses at Combs Lake. The water had receded to such a degree that more than two-thirds of the normal area of the lake consisted of shore mud and this was colonised, especially on the south side, by vast quantities of *Physcomitrium sphaericum* in association with *Physcomitrella patens, Pseudephemerum nitidum,* and *Ephemerum serratum.* During the late summer of 1949 a further period of drought occurred and even more shore mud was uncovered. A similar growth of annual mosses developed, though not in quite the same areas as in 1947, persisting until the water level began to rise again in late October. Toddbrook Reservoir was also visited on this occasion and *Physcomitrium sphaericum* was found to be present along with the same associates. During 1949 several hepatics which had not been observed during the 1947 season were found in the community. At Toddbrook *Blasia pusilla, Riccia sorocarpa,* and *Fossombronia wondraczekii* were noted in a number of places, while at Combs Lake there was widespread growth of *Riccia fluitans, Riccia glauca,* and an unidentified *Fossombronia.* The latter locality was made more remarkable by the presence of an abundance of the little flowering plant *Limosella aquatica,* which had not previously been recorded in the district.

Close observation of Combs Lake between 1950 and 1959 revealed very little of the bryophyte communities mentioned, though on one or two occasions when the water in the main channel feeding the lake was somewhat low, a small quantity of *Physcomitrium sphaericum* and its associates appeared on the marginal mud. The drought of 1960 once again produced conditions favourable to the growth of these plants, with the result that they were present in quantity from August until early in November.

LICHENS

Most upland regions throughout Great Britain are notable for the richness of their lichen flora, but in this respect the Peak District is somewhat disappointing. The relatively limited range of species and lack of luxuriance of growth in many cases is perhaps due in large measure to the atmospheric pollution from nearby industrial centres and from the quarrying and lime-burning both within and on the borders of the National Park. Lichens are probably more susceptible to the effects of smokeand fumes than any other type of plant and soon become impoverished and tend to die when the air becomes in any way impure.

The limitations of the Peakland lichen flora are most evident when one studies those which grow upon the bark of trees, for corticolous species are particularly scarce on the oak, which is the dominant tree on siliceous soils, and on the ash which occurs in the limestone dales. Though not abundant, such plants as *Evernia prunastri* and *Pamelia physodes* are characteristic of these habitats.

Lichens which grow on the face of bare rock – saxicolous species – are much better represented. Outcropping rocks are an integral part of the Peakland scenery both in the gritstone and limestone regions, and present extensive areas which can be colonised by these plants. Some species, such as *Parmelia saxatilis,* are associated solely with the gritstone, while others, like *Placynthium nigrum,* grow only on the limestone boulders, walls and cliff-faces. Such saxicolous species play a considerable part as pedogenic agents, helping to break up the rock and prepare it for plants more dependent on loose soil. In the case of those species confined to siliceous rocks the action is purely a physical one, but where the lichens grow on limestone the complex acids secreted by the thallus gradually dissolve away the rock and aid in its disintegration. In this way lichens are indispensable pioneers of vegetation.

The bare soil of moorlands as well as the grassy limestone banks in the dales provide a third broad habitat for lichens. Plants of such habitats are termed terricolous species and many of them are leafy in character rather than crustaceous. *Cetraria aculeata, Cladonia sylvatica,* and *Cladonia rangiformis* are characteristic moorland lichens, while *Peltigera canina* and *Solorina saccata* are widely distributed on limestone dale slopes.

The following list, though by no means complete, gives some idea of the Peak District lichen flora.

CORTICOLOUS LICHENS

Parmelia physodes
P. caperata
P. saxatilis
P. fuliginosa var. *laetevirens*
Evernia prunastri
Physcia pulverulenta

P. grisea
Lecanora conizaeoides
Pertusaria pertusa
Lecidea parasema
Biatora quernea (on *Prunus spinosa*)
Lecanactis abietina
Opegrapha rufescens
Graphis elegans
Acrocordia coniodea

SAXICOLOUS LICHENS

Gritstone

Parmelia saxatilis
Physcia caesia
Squamaria muralis
Cladonia flabelliformis
Biatora coarctata var. *elaschista*
Lecidea macrocarpa
L. cinereoatra

Limestone

Placynthium nigrum
Collema tunaeforme
C. tenax
C. granuliferum
Leptogium plicatile
L. schraderi
L. sinuatum (rock crevices)
Solorina saccata (rock crevices)
Porina chlorotica
Parmelia scortea
Caloplaca aurantia
C. luteoalba
C. rupestris
Physcia leptalea
P. virella
Rhinodina bischoffii
Lecanora albescens
Aspicilia (Lecanora) calcarea
Solenospora candicans
Cladonia pyxidata
C. pyxidata var. *pocillum*
C. cervicornis
Gyalecta jenensis
Lecidia jurana
Biatora immersa
Biatorella pruinosa
Thalloidima coeruleo-nigricans

Diploichia canescens
Dermatocarpon miniatum (rock crevices)
D. lachneum (rock crevices)
Verrucaria sphinctrina
V. viridula
V. nigrescens
V. muralis
Thelidium leightonii

TERRICOLOUS LICHENS

Peltigera canina
P. canina var. *rufescens*
Cetraria aculeata
Cladonia sylvatica
C. furcata
C. rangiformis
Biatora granulosa - on peat
B. uliginosa - on peat
Bilimbia lignaria

LICHENS FROM MISCELLANEOUS HABITATS

Xanthoria parietina - on old roofs, walls mainly of limestone and occasionally on trees
Diploschistes bryophilus - overspreading mosses
Cladonia ochrochlora - on rotting wood
C. flabelliformis - on rotting wood
Bilimbia sabuletorum - on mosses
Bacidia muscorum - on mosses
Botrydina vulgaris - on mosses
Verrucaria elaomelaena - on stones in streams
V. aethiobola - on stones in streams.

FUNGI

Unlike the plants which have been considered earlier in this work, fungi do not possess chlorophyll and are unable to synthesise carbohydrates from the carbon dioxide of the air and the water of the soil solution. This point does not mean that fungi are entirely devoid of pigments: some are brilliantly coloured like the scarlet toadstools of the Fly Agaric (*Amanita muscaria*[1]), others are more sedate, like the Blewit (*Tricholoma personatum*), which has fawn caps and purple blue streaked stems. The absence of chlorophyll imposes a special kind of nutrition upon fungi so that they depend upon organised sources of carbon for their energy supplies and thus exist as parasites in the bodies of living plants and animals or as saprophytes in dead remains of plants and animals. Therefore they are closely associated with the substrates from which their food materials

[1] Examples cited will be found in the list of species.

are derived and, to a degree, their distribution in nature corresponds with the distribution of suitable habitats and food sources. They are exceedingly numerous and many of them can be studied properly only with the aid of a microscope. The casual wayfarer in the Peak District will notice but a small number of the many hundreds of fungi which occur in this area. Mushrooms and toadstools become obvious at sporing time because of the relatively large size of the fruiting structures which emerge from the ground – diversely shaped umbrella-like toadstools such as Field Mushroom, Blewits, Death Cap; spherical or pear-shaped puff-balls (Lycoperdon and Bovista species); cylindrical and branched stags-horn-like structures (Clavaria and Xylaria species); or columnar and capitate phalloids (Stinkhorn). Logs, stumps and large branches of trees also carry their crop of toadstools *(Pluteus cervinus)* but more characteristically bracket-shaped and rather tough, leathery or woody, 'dryads-saddles', of many shapes, hues and sizes *(Polyporus squamosus: P. betulinus)*. The parts of these fungi which are visible are specialised for the production and liberation of spores; the active, and often very extensive, absorbing body of such fungi consists of multitudes of fine threadlike filaments within the matrix on which the fungus lives. These are seldom visible without microscopical examination, but in a specialised form, as the black bootlace-like rhizomorphs of Honey Agaric *(Armillaria mellea)*, they may be seen under the bark of dead logs or growing over the exposed wood of dead, standing trees.

Those who are interested in this very important group will discover a fund of information and interest in *Mushrooms and Toadstools* by Dr J. Ramsbottom, published in hardback in the New Naturalist Library in 1953, which is beautifully illustrated by photographs, many of them in colour. *Common Fungi* by Miss E. M. Wakefield contains descriptions of many of the usually encountered larger fungi and is illustrated with photographs and coloured plates of a wide range of species. For more advanced information, *Common British Fungi* by E. M. Wakefield and R. W. G. Dennis, should be consulted.

The following list includes only the commoner large fungi listed for the Peak District National Park. The habitat classes are indications of where these fungi may be sought.

WOODLANDS

Mixed Woods

Tricholoma nudum (Wood Blewits), also in compost and leaf mould

Clitocybe brumalis
Collybia radicata
C. maculata, also in bracken
Marasmius peronatus (Wood Woolly-foot)
Mycena pura, also leaf mould
Omphalia fibula, in damp moss
Inocybe geophylla
I. asterospora, in damp places
Naucoria cucumis, wet places: leaf piles
Flammula tricholoma
Tubaria furfuracea, and at roadsides
Paxillus involutus
Psalliota sylvatica (Brown Wood Mushroom)
P. haemorrhoidaria (Bleeding Mushroom)
Stropharia aeruginosa, and in pastures
Coprinus atramentarius (Ink Cap), and in pastures
Cratarellus cornucopoides
Boletus edulis
B. reticulatus
Clavaria cinerea
Phallus impudicus (Stinkhorn)

Deciduous Woods
Clitocybe odora
Marasmius erythropus
Lactarius blennius
Russula nigricans
*Cantharellus cibarius** [1](Chanterelle)
(a) Mainly associated with beech
Armillaria mucida, on trunks and branches
Collybia fusipes, also associated with oaks
Mycena capillaris, leaves
Russula azurea
R. cyanoxantha
Cortinarius anomalus – and mixed woods
Inocybe fastigiata
(b) Mainly associated with birch
Amanita muscaria (Fly Agaric) – and with pine
Cortinarius violaceus – and with pine
C. armillatus
C. hemitrichus – and heaths
Polyporus betulinus
(c) Mainly associated with oak
Lactarius quietus
Cortinarius hinnulus
(d) Mainly associated with alder
Naucoria escharoides – in damp sites

[1] Edible species indicated by asterisk.

Coniferous Woods

Lactarius deliciosus
Russula emetica – and deciduous
Hebeloma mesophaeum
Boletus badius
Fomes annosus

Mainly associated with pine

Clitocybe aurantiaca – and heaths
Collybia clusilis – and heather
Marasmius esculentus – buried cones
Hygrophorus hypothejus – and heaths
Cortinarius cinnamomeus
C. obtusus – with moss
Flammula sapinea – chips: branches
Gomphidus viscidus
Clavaria flaccida

Standing Trees

Polyporus squamosus – elm mainly
Fomes ribis – hawthorn: currants
Auricularia auricula-judae – elder

Logs and Stumps

Tricholoma rutilans – pine
Armillaria mellea – and base of old trees
Mycena gallericulata
M. polygramma – and on ground
*Pleurotus ostreatus** (Oyster Fungus)
Pluteus cervinus
Pholiota squarrosa – and dead deciduous trees
Crepidotus mollis
Hypholoma fasciculare
Psathyrella disseminata – and ground
Coprinus micaceus
Fistulina hepatica – oak, sweet chestnut
Polystictus versicolor
P. abietinus – and branches of conifers
Stereum hirsutum
S. purpureum
S. gausapatum – oak
Tremella mesenterica – deciduous
Calocera viscosa – conifers
Dacromyces deliquescens – and worked timber
Lycoperdon pyriforme

Woods and Heaths

Amanitopsis vaginata (Grisette)
Clitocybe nebularis
Laccaria laccata

Hebeloma crustuliniforme
Omphalia umbellifera – peaty
Galera hypnorum – in moss
Schleroderma aurantium

PASTURES AND GRASSLANDS

Pastures

*Tricholoma personatum** (Blewits) – and downs
*T. gambosum** (St George's Mushroom) – calcareous
*Marasmius oreades** (Fairy Ring Champignon) – lawn
Hygrophorus virgineus – and downs
T. coccineus – and downs
Entoloma porphyrophaeum
E. sericeum – and lawns
Eccilia undata – and on dead leaves
*Psalliota campestris** (Field Mushroom)
P. arvensis (Horse Mushroom)
Paneolus papillionaceus

Woodland Glades and Pastures

Amanita phalloides (Death Cup)
A. rubescens ('The Blusher') – and heaths
Lepiota procera (Parasol Mushroom) – soils
Nolanea staurospora
Clitopilus prunulus
Galera tenera – and roadsides
Lycoperdon perlatum

Pastures and Heaths

Hygrophorus pratensis
Leptonia lampropus
Stropharia inuncta
Bovista nigrescens

Roadsides

Naucoria semiorbicularis – and pastures
Hypholoma velutinum – and pastures
Psilocybe semilanceata – and heaths
Psathyrella gracilis – and woodland paths
P. atomata – and fields

VEGETABLE DEBRIS

Tubaria inquilina – twigs and dead grass
Crepidotus variabilis
Crucibulum vulgare – twigs: fern fronds

COMPOST AND HUMUS SOILS

*Lepiota rhacodes** (Shaggy Parasol mushroom)

*Coprinus comatus** (Shaggy Cap or Lawyer's Wig) – and waste tips
Peziza aurantia (Orange-peel Elf cup)

DUNG

Bolbitius vitellinus
Stropharia semiglobata
Anellaria separata
Paneolus sphinctrinus
P. campanulatus
Coprinus fimetarius

Fungi parasitic upon plants are almost all microscopic but some make themselves known by the reaction of their hosts to infection and at the time of sporulation by the colour of their masses of spores. This is true of the rust fungi which are well represented in the Peak District particularly in the more humid dales where lush vegetation provides an excellent habitat for their existence. Spring and early summer are the best times to search for such fungi, and one of the earliest and easiest to find is *Uromyces Ficariae* on the leaves of *Ficaria verna* (the Lesser Celandine). This fungus produces yellowish lesions on leaves and petioles, and at sporing time dark brown to black masses of spores break through the infected areas, which are swollen and distorted in such a marked manner that they are quite obvious when walking through a patch of Lesser Celandine. About the same time another *Uromyces, U. Poae,* infects the same host but produces 'cluster-cups' upon leaves and petioles. These appear as numerous minute cups filled with dust-like masses of yellow spores and under a lens the torn, recurved margins of the cups can easily be seen. This fungus requires one of several species of Meadow Grass to complete its life-cycle, and it is on these, in the late summer or autumn, that the black spore masses corresponding to those of *Uromyces Ficariae* are formed. Both hosts occur in close proximity in many of the dales. Another rust fungus growing upon two hosts is *Puccinia Caricis,* which in spring and early summer develops its 'cluster-cups' on very obvious, swollen and distorted areas of the leaves and petioles of the stinging nettle. These only occur when certain species of Sedge *(Carex)* are present in the vicinity to provide the alternate hosts on which, later in the year, long streaks of yellowish and black spores are produced upon the leaves and between the veins. In Dovedale and many other dales both hosts occur together and the fungus is found year by year in the same areas. On Meadow Sweet *(Filipendula ulmaria)* in many of the dales a very beautiful rust fungus *(Triphragmium ulmariae)* produces bright orange pustules on distorted areas over the veins of infected leaves or on swollen areas of petioles. In the autumn other, darker or black, spore pustules are formed by this fungus either on the spots of original infections or on fresh leaves of the same host.

Other simpler and less obvious fungi are to be found on many of the dale-land plants. Only one of these will be mentioned because it is relatively easy to find. Dog's Mercury *(Mercurialis perennis)* is common on many of the dale slopes (e.g. Dovedale), and in spring and early summer certain plants are a lighter green and often smaller than their neighbours. If such plants are closely examined they will be seen to bear glistening, blister-like swellings due to infection by *Synchytrium mercurialis.* This is an extremely minute parasite which can only be observed in detail in a section of the infection under a microscope, but it is a relative of the organism responsible for 'Wart-disease' of potato, a trouble now infinitely less common than it used to be.

Amongst saprophytic fungi, other interesting associations with particular host species are to be found in the group known to mycologists as the Pyrenomycetes because of the often black and almost always carbonaceous nature of their fruiting structures. The majority of them live upon dead plant materials. Many of them are extremely small and require microscopic study for their identification, but the Candle-snuff Fungus *(Xylaria hypoxylon)* should be familiar to many ramblers. It occurs on decaying logs, branches, and stumps at or close to ground level, and forms black stag-horn-like branches tipped with white waxy spore deposits in early spring and summer. This fungus is ubiquitous and not selective of the substrates on which it lives. But, *Daldinia concentrica* is almost confined to large logs and stumps of Ash, and is to be found in many of the dales. The sporophores are hemispherical and up to three or more inches in diameter, at first dark chocolate brown they later become black and, if cut vertically with a sharp knife, show marked concentric zonation of the inner tissues. Close relatives of this fungus occur on other trees. *Hypoxylon coccineum* on beech logs and branches has smaller dome-shaped, dull red sporophores, usually occurring in groups emergent from the bark surfaces. *Hypoxylon fuscum* has even smaller, flatter, hemispherical sporophores which are dull brown and finally black on Hazel, and richer brown, slowly becoming black, on Alder. This fungus is usually found on dead branches still attached to the bush or tree and is confined to these two hosts. On Hazel, Birch, Alder and Beech another type of pyrenomycete, *Diatrypella favacea,* produces black, emergent, rough, discoid or irregular pustules which look like coarse vegetable graters and cover considerable areas of the dead branches either while still attached to the tree or when lying on the ground. Smaller, regularly disc-shaped pustules belonging to the related *Diatrype disciformis* are confined to branches of Beech where the fungus may cover extensive lengths of freshly fallen branches. These discs are smooth to the touch and resemble minute, black coins scattered closely over the bark. *Diatrype stigma* forms extensive areas of black fungal tissues beneath the bark of branches of Oak, Beech, Hawthorn, Hazel, and certain other trees, and at sporing time bursts through the bark, which rolls back in a characteristic manner to expose the black crusts, often fissured transversely, and covered with minute punctate openings through which the spores are shed.

In Ilam Hall grounds and in certain of the dales another member of this group is to be found mainly on Box, but occasionally on other hosts. It is known to most British mycologists as *Nummularia lutea* and it completely decays the wood of its host, staining it bright yellow. One way of finding the fungus is to kick decayed Box wood: if it is yellow and very rotten then the fungus is almost certain to be found sporing on the rather harder external surfaces. Here it forms grey slightly concave discs of variable size which just emerge from the wood surface or the bark if any still remains. In surface view these sporing structures look very like small coins and the first name of the fungus refers to this similarity.

This very brief survey of some of the fungi occurring in the Peak District has sought to emphasise the fact that these organisms are governed in their distribution by their habitat conditions and particularly by the availability of substrates containing the necessary food materials to support their growth. In this National Park a wide diversity of fungi exist and very few of them have been recorded or studied in any detail. A field of interest and value awaits anyone keen enough to tackle problems related to the identification and distribution of this group of organisms.

REFERENCES

Linton, W. R. *Flora of Derbyshire.* Bemrose (1903)
Report of the Brit. Bryological Soc. (1922)

Dixon, H. N. *Handbook of British Mosses.* Sumfield (1924)
MacVicar, S. M. *Handbook of British Hepaticae.* Sumfield (1926)
Verdoorn, J. C. *Manual of Bryology.* (1932)
Trans. of the Brit. Bryological Soc. (1947)
Watson, E. V. *British Mosses and Liverworts.* Cambridge University Press (1955)
Wakefield, E. M. and Dennis, R. W. G. *Common British Fungi.* Gawthorn (1950)
Ramsbottom, J. *Mushrooms and Toadstools.* Collins: New Naturalist (1953)
Wakefield, E. M. *Observers' Book of Common Fungi.* Warne (1954)
Harley, J. L. *Biology of Mycorrhiza.* Leonard Hill (1959)
Dennis, R. W. G. *British Cup Fungi and their Allies.* Ray Society (1960)

8. Early Man in the Peak

What a piece of work is a man!
William Shakespeare: Hamlet

Among the detailed features of the Peak scenery, rock stacks and pinnacles, and curiously-shaped boulders, often forming conspicuous landmarks, are a constant source of interest. At a distance it may be difficult to decide whether they are natural or man-made as in the case of the huge block on Alport Hill or the two pillars called Inaccessible and Weasel of Robin Hood's Stride on Harthill Moor. These of course are natural features, the results of weathering upon bare outcrops of gritstone. On the other hand, close scrutiny of the massive accumulation of rocks known as the Carl Wark on the flank of Higger Tor betrays unmistakably the impress of human labour, a crude fortress raised by early man. Even more striking is the stone circle of Arbor Low, which is far older. There are many other relics of the distant past in The Peak, not all of them so prominent, but they are of great interest since they are what remains of the first marks made by man upon the landscape at a time when his material progress was still slow and faltering.

THE PREHISTORIC LEGACY

Actually the Peak District, together with the lowland fringes to the east, south and west, provides a rich and eloquent record of prehistoric cultures from the Old Stone Age to the Iron Age. While it is true that only limited evidence is available concerning the movement of early peoples into Britain from outside, there is ample testimony to the continuity of culture and of its adaptation to changing conditions. It is in this respect that The Peak occupies a rather special position in archaeology. In contrast to the surrounding lowlands, it was penetrated only by relatively small numbers belonging to successive migrations which brought new cultures to these islands. As a result each new culture, reaching the uplands by tenuous lines of movement, was largely absorbed and modified by the existing one. To this fact must be attributed a certain lack of clear differentiation between the characteristic features of one cultural phase and the next.

In Britain and north-west Europe primitive man belonging to the Old Stone Age first appeared during a warmer interval in the Great Ice Age and gained a livelihood by hunting. Not far to the east of the Peak District the caves at Creswell have yielded a full and unique record of his sojourn in that part of the country. In striking contrast are the few finds belonging to that time which have been made within The Peak. These consist of a few quartzite tools from a cave at Harborough Rocks near Brassington and a flint from the Ravenscliff Cave in Cressbrook Dale. A black chert hand-axe of Old Stone Age pattern, found in the gravels of the Trent, points to the early use of this material which occurs in the limestone of The Peak. These finds, however, are sufficient to indicate that Old Stone Age or Palaeolithic man at least ventured into the area.

Evidence is much more abundant for the presence of prehistoric man in post-glacial times. A good introduction to this part of the story is furnished by discoveries made in the peat cover of the moorlands, especially on Warwick Hill near Marsden. Here under a layer of peat three feet thick the surface of the gritstone is covered by an ancient sandy soil in which flint flakes have been found exhibiting the workmanship of Mesolithic or Middle Stone Age man who lived in Britain possibly from 5000 to 4000 BC. These flints were found in such numbers as to suggest that man was at that time more than a mere visitor to the Peak District. Plant remains found in association with these relics show that even at these high altitudes birch-oak woodlands flourished and that the climate was sufficiently genial for him to live in the open air.

Fewer traces of Mesolithic man have been found on the limestone, the only implements so far discovered having come from Thor's Cave in the Manifold Valley and from Demon Dale near Taddington. Whether on the whole man avoided the limestone country at this time it is impossible to say, for as the late Mr A. L. Armstrong showed, the frequency of implements from the sandy gritstone may simply indicate the readiness with which wind erosion exposes them and not necessarily man's preference for such areas.

NEOLITHIC PASTORALISTS

Towards the close of this period the climate, though still warm, became moister and consequently peat began to form on the moorlands. This eventually cloaked the ground so effectively that roots were smothered, trees died and the woodlands were extensively destroyed. On Warwick Hill, already mentioned, the presence of leaf-shaped arrow-heads in the bottom five inches of the peat shows that New Stone Age or Neolithic man had come into the district. The discovery in the next two inches of a piece of bronze with arrow-heads of Bronze Age type, and of fragments of Romano-British pottery in the middle layers of the peat, provide glimpses of the later stages in the human occupation of The Peak. The remaining eighteen inches of peat have accumulated since the Roman era in Britain.

With so genial a climate it is not surprising that early man was much less dependent upon caves for shelter and that these have yielded disappointingly few relics of his presence in them. These limited discoveries included a lump of pyrites found in the Longcliffe Cave. This was probably used for lighting fires by striking sparks as from flint. Ravenscliff Cave, already mentioned, has yielded pottery made by hand and a bone tool that was used in the process. From Cave Dale near Castleton have come crude pottery, a bronze axe and a perforated stone hammer. Evidently both New Stone Age man and Bronze Age man made use of these caves.

The New Stone Age was initiated in the Peak District by immigrants from the west who appear to have entered the uplands in the neighbourhood of where Macclesfield now stands. The newcomers were herdsmen and the success of the new means of livelihood based on stock-breeding and grazing appears to have encouraged a good deal of racial fusion with the surviving Mesolithic inhabitants. Other Neolithic innovations also made their mark, especially the erection of large stone monuments and the use of stone for burial chambers, relics of which are widely scattered over the area. Because the Neolithic people used large stones for such purposes they are also known as the Megalithic folk.

The largest and most striking of these monuments is the stone circle at Arbor Low. This is one of the three largest late Neolithic or early Bronze Age stone circles in England, though it is less spectacular than either Stonehenge or Avebury. Surrounding it is a circular mound 50 feet above the general level of the ground. Within this is a ditch which was originally 10 feet deep. This defines the limit of a flat platform upon which is a circle of over forty large slabs of limestone lying upon the ground and pointing towards the centre. At this point are several larger stones, the remains of some previous structure. Gaps in the mound, coincident with filled-up portions of the ditch, mark the positions of two entrances at the N.N.E. and S.S.W. points respectively. The purpose of all this is unknown. It can only be surmised that it was the scene of some ritual which took place on important occasions when possibly patriarchal groups from far and near assembled and looked on with awe from the mound which served as their grandstand. Such a monument also suggests that late Neolithic communities lived an organised life with an established social and economic order of which Arbor Low can be taken as a symbol. Arbor Low is certainly the best known prehistoric site in The Peak, and its position at an altitude of 1,200 feet, giving impressive views over the limestone country, is hardly surpassed by Salisbury Plain. Many tumuli are to be found in the neighbourhood, built presumably so that the dead might repose near the great monument. The name Low, which abounds in The Peak, means mound, burial mound, or hill.

Except for the absence of stones, the Bull Ring at Dove Holes is a comparable monument. Whether the stones were there originally or whether the work of construction was cut short by the arrival of other people with a different culture is not known.

The majority of the other Megalithic relics are associated with the disposal of the dead. Using large slabs, one or more chambers, sometimes several, were made for burial purposes. After laying a paved passage-way to these chambers, the whole was covered with a large mound, leaving an entrance to the passage at one side. About a dozen of these chambered barrows are known to have existed in the Peak District some of which were sufficiently well preserved to illustrate the method of construction just outlined and to yield a number of skeletons. The latter showed that Neolithic man was moderately tall and that his head was slightly elongated from back to front. The barrows at Five Wells near Taddington, Minning Low and Harborough Rocks deserve special mention, but others, not so well preserved, occur at the Bridestones near Congleton, Green Low, Long Low and Ringham Low. Five Wells, standing at 1,400 feet, is the highest Megalithic structure in England.

Besides these chambered tombs for collective burial, many smaller barrows were erected over cavities cut out of the rock or built of limestone slabs as at Hob's House, Monsal Dale.

These served as burial-places for individuals, the body being interred in a crouched position.

The evidence of hand-made pottery of a kind called the Peterborough type shows that another branch of the Neolithic culture had come into the Peak District from the east. Fragments of such pottery have been found in places as far into the interior as the High Wheeldon Cave near Earl Sterndale and in a barrow near Arbor Low. Thus, considering all the relics so far brought to light it would seem that the limestone country in particular was relatively well settled by Neolithic man. There are also traces of a system of trackways by means of which the exchange of products took place from one area to another. Yet on the higher northern moors some Mesolithic hunting communities lingered on almost unaffected by the material advances taking place to the south.

BRONZE AGE PEOPLE

Between 2000 and 1500 BC another race bringing the Bronze Age culture entered the district from the east. Physically they were of medium height with relatively broad skulls. In their tombs, however, long-headed types occur in sufficient numbers to indicate that the new arrivals and old inhabitants settled peaceably together. The burial customs altered, for chambered tombs were no longer used. With the body were placed drinking vessels or beakers. Later these were replaced by food vessels of coarser workmanship. This change serves to distinguish two phases of the early Bronze Age culture, those of the Beaker people and Food Vessel people respectively. Over the burial-place a mound of earth was heaped, forming a round barrow. Of the round barrows which have yielded beakers, well over twenty are located on the limestone country, whereas they have proved rare on the gritstone moors.

The Food Vessel people were probably the first to use bronze implements, for numerous examples of food vessels have been found along with polished flint axes and bronze weapons. Since nearly all the bronze objects are known to be of Irish origin there must have been considerable trade with the inhabitants of the Low Peak in particular. It is evident that some of the tribal chiefs were wealthy enough to provide their womenfolk with ornaments of Yorkshire jet. Though the pastoral economy continued, the cultivation of grain crops began, and the need for more land for this purpose is reflected in the use of heavy stone axes for removing part of the forest.

Towards 1000 BC a fresh influx of Bronze Age people introduced the custom of cremation. In this case the ashes of the dead were placed in a small cist made of stones or covered with an urn turned upside down. Quite often they were placed inside the urn and covered with a stone. In any case the ashes were buried under a mound of earth. Sometimes these cinerary urns were deposited in clusters to form cemeteries. One of these cemeteries on Stanton Moor yielded fifteen urns. From this custom these newcomers are known as the Urn people. With superior weapons, including the spear, they appear to have subjugated the Food Vessel communities and gained control over much of the Peak District.

Besides settling in the limestone area, the Urn people penetrated the gritstone moors, where small stone circles or even solitary stones were erected. Examples of these are the Nine Ladies on Stanton Moor, the four remaining stones of the Nine Men of Harthill Moor, together with the stone circles on Froggatt Edge, Abney Moor, Moscar Moor and Ramsley Moor. These sites indicate not only an increase in population, which must have been accompanied by further forest clearing, but a substantial spread of settlement to the higher gritstone country.

The Bronze Age folk were just as highly skilled as their Neolithic predecessors in the working of flint. Their arrow-heads had a barb and two tangs. They also made flint knives and notched one edge of similarly shaped flakes of flint for use in making a cutting edge for their wooden sickles. They made beautifully shaped stone hammers through which they bored holes into which a haft was fitted. They cast palstaves in bronze and used them for breaking up and hoeing the soil. The relics found in their burial-places include daggers, pins, buttons and other clothing accessories, from which it is inferred that they buried or burned their dead fully clothed. All this points to the existence in the Peak District of simple agricultural communities enjoying some of the amenities of a primitive civilisation.

THE IRON AGE

The earliest phase of the Iron Age in Britain dates from the fifth century BC, but of the first people to enter from the Continent possessing a knowledge of iron, the Peak District bears hardly a trace. Only belatedly did the Bronze Age culture give way. One distinctive feature of the Iron Age, however, gradually appeared. This was the hill fort with its earth rampart and ditch generally constructed on the summit of a commanding hill. Examples are to be found on Mam Tor, Fin Cop, Ballcross near Bakewell, and Combs Moss, south of Chapel-en-le-Frith, while on the south-east shoulder of Higger Tor is the massive structure called the Carl Wark. These forts are of varying date but were in use for a considerable period, some of them well into Roman times. As a rule they are smaller than those of Wessex, yet in their rather sombre surroundings they are equally impressive. The Carl Wark, with its rampart faced with large gritstone blocks forming a wall nearly ten feet high, is probably of relatively late construction. It was long supposed to be of Celtic Iron Age date but is now regarded as post-Roman. By the time of the Roman invasion the older hill forts ceased to be of strategic value and were used only as refuges in local warfare or as shelters against cattle-raiders.

Another relic of the Iron Age is the quern used for milling corn. Many of these have been found even on the northern moors, indicating that pastoral farming, though still the basic activity, was supplemented by cultivation. This was in fact the situation at the coming of the Romans. Their impact upon The Peak during the first century AD, especially during the governorship of Agricola, was chiefly to establish a system of roads and forts with which to complete the subjugation of the southern Pennines, a process which broadly speaking marks the division between prehistoric times and those of recorded history.

REFERENCES

Daniel, G. E. *The Prehistoric Chamber Tombs of England and Wales.* Cambridge University Press (1950)

Preston, F. L. Hill Forts of The Peak. *Jour. Derbys. Arch. and Nat. Hist. Soc.: LXXIV* (1954)

Piggott, S. *Neolithic Cultures of the British Isles.* Cambridge University Press (1954)

Fowler, J. The Transition from Late Neolithic to Early Bronze Age in the Peak District of Derbyshire and Staffordshire. *Jour. Derbys. Arch. and Nat. Hist. Soc.: LXXV* (1955)

Armstrong, A. L. Section 6, Prehistory (Palaeolithic, Neolithic and Bronze Ages) from *Sheffield and its Region.* British Association for the Advancement of Science (1956)

Bartlett, J. E. Section 6, Prehistory (Iron Age and Roman Period). *Ibid.*

Thompson, D. *Guide to Arbor Low.* H.M.S.O. (1963)

9. Villages and Farms

> In every village marked with little spire
> Embowered in trees, and hardly known to fame.
>
> *William Shenstone: The Schoolmistress*

The development of human settlement in the Peak District presents as fascinating a story as that recording the evolution of the scenery. From far-off prehistoric times until the present, man has continuously been able to find in this hilly country the means of livelihood by which to maintain his existence. Over much of the region he has always had to contend with conditions of surface relief and climate less favourable than those of the lowlands, but by adapting himself to these conditions he learnt at an early date to exploit the natural resources, to develop the techniques of food production and to organise his life on a permanent basis.

Through such a long span of time the pattern of settlement has naturally developed in stages as we have already seen in the case of some of the later prehistoric periods. At the present time the settlements take the form of numerous villages, together with a considerable sprinkling of dispersed farms and a few small towns. The villages are the characteristic units over much of The Peak. In their appearance and setting most of them give a pleasing impression and fit harmoniously into the landscape. They represent a system of settlement which is well over a thousand years old, for with few exceptions the beginnings of the present-day villages date from the period of the Anglo-Saxon occupation. Some Romano-British communities probably remained long after the Roman withdrawal and even after the first arrival of Anglian tribesmen in the English lowlands. Possibly the earthwork at Bradwell known as the Gray Ditch is the relic of a defence line against Anglian settlers advancing into the hill country. By the middle of the seventh century, however, Anglian settlement was certainly established in The Peak. Anglo-Saxon *-inga* names like Hartington and Taddington indicate an occupation dating from a little later than the first phase of colonisation. Forest-clearing by the Anglo-Saxons in some of the chief valleys is plainly demonstrated by the *leah* element in place-names along the Derwent and its tributaries: Wensley, Darley, Rowsley, Beeley, Padley.

There followed a phase of consolidation accompanied by the founding of more villages, except in the highest parts, which still formed a bleak frontier zone between the kingdoms of Mercia and Northumbria. A comparatively late penetration brought settlement into the Hope Valley, and by the concluding stages of the Anglo-Saxon colonisation, a great number of villages had been established not only in the principal valleys but over the limestone plateau itself, with the surrounding land cleared for the open-field system of cultivation. Most of the higher gritstone moors, however, remained unpopulated as they are today.

Place-name evidence shows that Scandinavian settlers did not penetrate westwards as far as the Derwent. They therefore made no contribution to the Peak settlement pattern, though a vague tradition affirms that the Danes worked lead from the Odin Mine near Castleton. Thus in general the distribution of settlements in the region was completed by the end of the Anglo-Saxon period, a fact which is confirmed by the record of the Domesday Survey of 1086. Since then it has undergone little change but for the addition of individual farms such as the monastic granges of medieval date and those resulting from enclosure in more modern times.

THE VILLAGES

The different forms assumed by the villages of The Peak are closely related to site and early function. Of the so-called clustered or compact villages of the limestone country, the commonest is the linear type, built along the narrow floor of a dry valley or on a bench above the valley such as Over Haddon, overlooking Lathkill Dale. In this type the farmhouses, separated by small closes, are arranged along both sides of the street. Later growth by infilling preserves the simple plan, merely making it more compact and converting it to a true street-village as at Taddington, Elton, Wensley and Eyam. Chelmorton, perhaps the best instance of all, shows a clearly marked pattern of long narrow fields separated by stone walls running at right angles to the street, indicative of the old medieval cultivation strips. Flagg and Wardlow are much more attenuated because of the exceptional spacing of the farms along the road. On the other hand, growth promoted by lead-mining in the late eighteenth or early nineteenth centuries resulted in the extension of many villages on to sharply rising ground, producing branch roads (often called 'banks') leading off the main street. Such development results in a less regular plan as in the cases of Bradwell, Stony Middleton, Brassington, Winster and Youlgreave. At Tideswell the expansion of an originally close linear settlement in a dry valley floor has given rise to parallel development on benches along both sides.

Distinct from the linear form is the village which grew around a green containing a pond for watering cattle. Though still compact, this type is irregular in plan, as can be seen at Foolow, Hartington, Monyash, Litton and Parwich.

The settlement pattern on the limestone differs in detail from that of the gritstone country. Villages predominate on the limestone and dispersed settlement is largely of a secondary nature, whereas on the grit and shale areas, except for the shales of the Low Peak, settlement is characteristically dispersed. Over the permeable surface of the limestone a close relationship exists between village site and water supply, for the latter is particularly local in occurrence. The permanent streams flow in deep dales somewhat difficult of access and divorced from the agricultural land which lies above the dales and on the plateau proper. Along the floor of the dales, therefore, few villages are found, and only in the case of the larger streams do normal valley settlements occur as at Alport, Ashford and Bakewell. Instead the villages grew up at various alternative sites: along the junction of the limestone and the shales, such as Great Hucklow and Great Longstone; on dale slopes or benches where the water-table is accessible by shallow wells as at Tideswell and Over Haddon; and above all, on the plateau itself where wells may tap supplies from fissures in the limestone or reach a perched water-table resulting from an occurrence of igneous rock. Some of the earliest villages were founded in the shallower dry valleys where not only was underground water obtainable but shelter and deep soil as well.

On the gritstone and shale country water is abundant at all levels, giving greater freedom of choice both for the location and the form of settlements. Hill-top, hill-slope and valley-bottom sites for farmsteads are all common, although a mid-slope position near a small stream is perhaps the most typical. The last-named may have arisen because of the need for access to both the improved pasture below the farm and the upland grazing above; elsewhere it appears to be related to aspect.

In those instances where they can be approximately dated, the oldest farms in Edale, the Woodlands Valley and the Derwent Valley are situated conspicuously on the south and south-west facing slopes.

In the higher gritstone country, the bleak moors and rugged 'edges' encouraged the development of valley settlements, including both villages and dispersed farms, as can be seen in Edale and the Derwent valley. Later on the use of streams for power and processing during early industrial times led to the erection of textile mills, which resulted in the further growth of some of these villages, especially along the Derwent, following Arkwright's successful venture at Cromford in 1771.

Of the Derwent villages, Baslow and Edensor deserve special mention for both have been profoundly affected by the establishment of Chatsworth House and the great park which surrounds it. Baslow is now a large village loosely composed of three parts called Bridge End, Nether End and Over End respectively. Far End, consisting of a group of farm buildings on the Sheffield road, is quite separate and distant from the village. Bridge End, with the church and the old bridge over the Derwent, was the original settlement. Developments at Chatsworth in the late eighteenth century, however, led to the growth of Nether End nearly half a mile away on the slope of the Bar Brook stream, which became the boundary of the park. A little later, in 1826, road improvements, including the diversion of the Chesterfield road to Chatsworth, made Nether End not only a centre for visitors to different parts of the Devonshire estates but the northern entry to the park itself. Hotels and residential development mark this phase of Baslow's growth, although the pleasant green which serves as a focus around which most of the older buildings are placed gives Nether End the appearance of being an entirely separate village. Residential development on the sunny slope above old Baslow, dating from the late nineteenth century, forms the district called Over End, 'over' being an Old English term for a bank or slope. The popularity of Baslow as a small resort and tourist centre, largely on account of its proximity to Sheffield and Chesterfield, has caused an immense increase in road traffic. For this reason the modern Derwent Bridge was erected, thus relieving the former congestion at Bridge End. Happily the old bridge remains and can be seen from the new, which is placed at a discreet distance from it.

Edensor, on the western edge of Chatsworth Park, is a curiosity among villages. It existed before the Norman Conquest and eventually developed as a typical agricultural settlement in the Derwent Valley. The village was situated on the right bank of the stream and extended up the slope to the west, while part of it spread across the Derwent towards the site now occupied by Chatsworth House. In the eighteenth century the fourth Duke of Devonshire caused the removal of those parts close to the river because they stood in full view of the great mansion. Some of the inhabitants found homes in the upper portion of the village, which was screened from the House by a spur of rising ground, while others were moved to the neighbouring village of Pilsley. The upper portion of Edensor itself was subsequently demolished by the sixth Duke, about 1839-40, transformed into a model estate village and rebuilt in the strange mixture of architectural styles which can be seen today. Individual houses in Gothic, Tudor, Renaissance, Swiss and Italianate styles present a freakish composition in stone, which time and the splendour of surrounding trees have done something to mellow. The church was also rebuilt and completed in 1870. A few old houses of the original village remain and the course of the former High Street can be plainly seen.

In comparatively recent years, largely as the outcome of railway and road transport, a new type of settlement has appeared in the Peak District. This is not a form of agricultural settlement at all but is really suburban in character. The opening of the railway from Sheffield to Manchester in 1894 brought the first-named city within easy reach of the Derwent Valley and the approaches to the High Peak. The line itself emerged from the darkness of the Dore and Totley tunnel upon the lovely stretch of valley between Grindleford and Hathersage, so different from the heart of Sheffield only ten miles away. Soon people began to build their homes around Grindleford, Padley and Hathersage, travelling daily to the city for their business. During the past forty years this development has been further extended by the use of the motor car. Residential building has spread along the slopes beneath Froggatt Edge in the one direction and from Hathersage towards Bamford in the other. These settlements have taken a linear form where they were related to the railway but have assumed a much looser pattern where they have resulted from road communication, thus reflecting the greater flexibility of that form of transport. The contrast between the rows of villas at Nether Padley and the scattered housing at Froggatt amply illustrates the difference. In the Padley-Grindleford-Froggatt district those who work at a distance now form almost the dominant element in the community. These commuters amount to over one-third of the employed population, the great majority going to Sheffield. Increasingly Bakewell provides houses for Sheffield workers.

The expansion of these villages in the Derwent Valley calls to mind the sharp distinction between the Derwent and the Wye in relation to settlements. Above the confluence point at Rowsley the valley of the former is carved in the grit shales and is relatively broad, while that of the latter, etched into the limestone, is narrow, and above Ashford-in-the-Water, takes the form of a winding gorge. Apart from Ashford itself there are no villages along the Wye between Bakewell and Buxton. Miller's Dale is really only a hamlet. Along the Derwent, however, there is a succession of villages, some on one side of the river and some on the other, extending from Beeley upstream to Bamford, and formerly up to Derwent until it was submerged beneath the Ladybower Reservoir. It is almost entirely the physical character of the two valleys which accounts for this contrast. The difference is further emphasised by the fact that in the limestone area the villages are found on the plateau high above the Wye, such as Great Longstone, Litton and Wormhill to the north of the valley and Sheldon, Taddington and Blackwell to the south, whereas on the East Moor, overlooking the Derwent, no villages occur at all.

The Derwent villages have actually experienced two phases of growth since the days when they were purely agricultural settlements. The first phase occurred in the early nineteenth century when the erection of textile mills using the river as a source of power gave added scope for employment. Even this modest degree of industrialisation was reflected by an increase in population. The second phase was that of the residential or dormitory development mentioned above, and this in particular has promoted a striking population growth. As already stated, the Sheffield-Manchester railway was opened in 1894. During the period of sixty years between the census of 1891 and that of 1951, the population of Nether Padley increased by over 300 per cent and that of Froggatt by over 100 per cent. Hathersage grew by 30 per cent and Baslow by 35 per cent. Bamford grew by well over 200 per cent, but this increase was due not so much to dormitory development as to the expansion of employment opportunities in the Hope Valley, to the housing of workers connected with the great reservoirs and to the accommodation of people displaced by the loss of Derwent village. In contrast to this, the population of both Ashford and Rowsley in the Wye Valley declined over the same period.

DISPERSED SETTLEMENTS

Turning to the dispersed settlements in The Peak, the great majority consist of individual farms. Since most of them were established at different periods long after the villages, they form a later or secondary element in the settlement pattern. A few on the limestone may be of Anglo-Saxon date and therefore almost as old as the villages themselves. Quite numerous, especially on the limestone, are the 'granges' which were founded by monastic institutions. The term 'grange' originally meant a granary. These farms were generally worked by lay brothers and provided revenue for the parent houses from the sale of wool. Many have remained as farms on the same site to this day even though their buildings have long since been replaced by others. Their names, containing '-grange' as a suffix, serve in most cases to distinguish them from later farms. Thus Griffe Grange near the head of the Via Gellia, was the property of Dale Abbey. Shortly before the Dissolution it passed to the Gell family of Hopton and remained in their possession until 1842. Similarly Mouldridge Grange was the property of Dunstable Priory. Cronkston, Griffe, Hanson, Meadow Place, One Ash, Royston and Smerrill Granges, among many others in the central area of the limestone, show that the religious foundations like the great flock-masters of Tudor times exploited the sheep-run country for wool. Dr G. J. Fuller has shown that within the ancient Hundred of Wirksworth there were no less than nine monastic granges in 1291 which were subject to Papal taxation. All but one of these were situated on the limestone. The wealthy Premonstratensian foundation of Welbeck in Nottinghamshire acquired considerable possessions in the Peak Forest, including a grange at Churchill, near Hope, two others in the Derwent Valley, and another in the Woodlands Valley from which the name Friars Wood has survived. From the celebrated Woolley manuscripts (thirteenth and fourteenth centuries) in the British Museum it is clear that several farms in the parish of Wirksworth originated as granges belonging to Darley Abbey. Among them, Aldwark Hough, Wigwell and Woodam remain today. Far to the north, near Chinley, is a small farm known as Monks Meadow which in all likelihood originated as a grange belonging to the distant Abbey of Basingwell, near Holywell in Flintshire. By no means have all the monastic farms persisted. Some for various reasons fell into ruins, while others mentioned in the cartularies have not been located at all, except that a place-name may occasionally provide a clue as in the case of Friars Wood and Monks Meadow already cited. On the other hand, not all farms bearing the name Grange today were of monastic origin, for a number have adopted the suffix more recently and without historical justification. To give one further illustration of the influence of the monasteries on the topography of The Peak, the Abbey of Lenton, near Nottingham, founded by William Peveril (who also built the stronghold at Castleton), was granted land by him near Tideswell. There, in what is still called Monk's Dale, a grange and chapel were established of which only the barest traces remain.

At different periods in the past the enclosure of waste land has been the means of promoting new farms. Such areas were appropriately called 'intakes' and Intake Farm, near Little Hucklow, may have originated in this way. Dr S. R. Eyre has shown that old field names bearing the element '-intake' occur frequently on the steep western flank of the East Moor and on the sharp slopes around the Ashover Valley. These areas, too, were carved from the waste, and in a few instances from woodland, and probably represent an encroachment upon land never previously cultivated which took place not long before the period of parliamentary enclosure. It should be pointed out that a good deal of the high ground on the northern gritstone outcrop as well as land on the eastern Coal Measures were never under the open-field system. As Dr C. S. Orwin has shown in his well-known work, *The Open Fields* (1938), much of this territory, including part of the King's Field was long under forest which fell into waste after it was denuded by generations of lead-miners. Of course the movement for Parliamentary Enclosure itself, which occurred relatively late in The Peak, gave rise to further individual holdings.

There is another feature of the settlement pattern in The Peak which deserves mention. This is the large house or hall, sometimes a manor, built long ago by well-to-do land-owners and members of the lesser aristocracy. There are many such houses in the region and most of them, though not of great size and splendour like Haddon or Chatsworth, are buildings of dignity and worth. Indeed their great number caused Phillip Kinder, writing about 1650, to remark 'No countie in England has so many princelie habitations.' They occur on the limestone and gritstone alike. Some stand adjacent to a village or within a village as at Tissington, others are more remote and are attached to farms. Most of them date from the late sixteenth and seventeenth centuries. They are built of gritstone in different versions of the Tudor and Jacobean styles, with thick walls, gables, mullioned windows with stone moulding and roofs of stone slabs. The weight of the roof in relation to the length of timber beams affected the planning of these houses, for very wide spans were impracticable. As a result they were uusually designed in the form of an L or E. One of the finest examples, and one of the oldest, is Aston Hall, an Elizabethan manor house situated on the lower slope of Win Hill. It dates from about 1580 and is now listed for protection as a building of architectural interest. Hartington Hall, though restored in 1862, belongs to the same period, while of rather later date are the Halls of Bakewell, Eyam, Hassop, Holme, Highlow and Offerton, to mention only a few. Derwent Hall, another fine example, was submerged in the Ladybower Reservoir with the last extension of the Derwent Valley Water scheme.

Having reviewed the essential elements in the settlement pattern, reference should be made to the distribution of the villages and farmsteads in relation to altitude since The Peak is an upland region imposing some restrictions upon the location of agricultural communities. In this respect further contrasts occur between the limestone and gritstone areas. On the limestone, villages occur predominantly in two zones of altitude, one at 600-650 feet above sea-level and the other at 900-1,000 feet. These zones of settlement broadly relate to the recognised erosion surfaces at 700 feet and 1,000 feet respectively which give rise to more or less level stretches of ground. These are the areas most suited to cultivation and naturally the settlements are closely associated with them, although as we have seen, the actual sites of villages are to be found a little below the plateau surface, on the marginal slopes or near the dale heads. Two exceptions are to be found in the upper zone where Chelmorton and Taddington, both situated at 1,100 feet, stand on the plateau itself. They are the highest villages in The Peak and are probably so placed because of their access to water. Despite the elevation and somewhat exposed position good crops of wheat can be raised in the neighbourhood as was proved during the last war.

In the case of the gritstone country where much of the high moorland provides poorer land for agricultural use, the majority of the settlements lie within a zone ranging from 400-700 feet. On favourable slopes individual farms may be found at 800 feet or even higher, but the chief exceptions as far as villages and hamlets are concerned occur towards the head of the larger valleys. Thus the hamlets of Upper Booth and Grindsbrook in Upper Edale lie at 900 feet and, until they were removed by the Stockport Corporation Water scheme, a few farms towards the head of the Goyt Valley were as high as 1,000 feet. Elsewhere Abney is at 978 feet.

THE USE OF STONE

The whole of the Peak upland is traditionally an area of stone buildings. Good stone can be obtained from many layers in the millstone grit and the limestone, but the latter, though generally as durable, is less easily dressed than the gritstone. The cost of transporting building materials in the past has resulted in a close dependence upon local stone for dwellings, farm buildings and the ubiquitous field walls. Even a minor inlier of limestone, as at Mixon near Leek, is reflected by limestone walls and buildings.

A preference for gritstone is often found, especially on the shales and in other places where both kinds of stone are accessible; on broad tracts of shales, however, limestone is used, as in the group of villages which include Tissington, Thorpe and Parwich, for the nearest gritstone quarries at Birchover are some ten miles distant. Ashford-in-the-Water is another village where many delightful cottages are built of limestone. Whether gritstone or limestone is used for building, the former is almost always preferred for doorposts, lintels, window mullions and gateposts.

Grit flagstones or similar thin flaggy sandstones from the Lower Coal Measures in the extreme east long provided a durable roofing material in The Peak. While these are seldom used today, they survive on all the older buildings and the characteristic arrangement by which the rows of narrower and lighter stones are placed nearest the ridge succeeded by those of increasing size and weight towards the eaves, can be seen everywhere.

Some houses in the limestone area are covered with a lime wash, which is also traditional, or a rough cast of limestone chippings, while a coating of 'pebble-dash' is a more recent feature. Such an addition is seldom found in gritstone settlements, and even where it does occur its purpose is usually to conceal brickwork and so to preserve harmony of appearance. Fidelity to local stone is greater in the older and smaller villages, for those which have grown in recent times like those in the Hope and Derwent Valleys, exhibit an unfortunate variety of materials, including brick, concrete and corrugated iron. In the large dormitory villages buildings of modern brick are made the more incongruous because of their alien design. At Nether Padley a recent survey showed that one-fifth of all the buildings are now of brick and tile, presenting an incongruous mixture of non-traditional styles. Brick and tile buildings are not uncommon in the extreme south as at Ashbourne, for the red marls of the South Derbyshire Trias, providing the raw material, lie close at hand.

The traditional dwelling-house in The Peak is a two-storey building of rather narrow roof-span with gable ends and little in the way of ornamentation. In the villages the customary grouping of houses and farm buildings into rows preesents low horizontal lines generally harmonising with the site. This impression of horizontality is often emphasised by the walls which divide the fields, although any tendency to monotony is averted by the clumps of trees which are a feature of most villages and almost all the solitary farms. Happily there is now a revival of the use of stone in domestic building within the National Park, generally with very pleasing results.

STONE WALLS

The most characteristic feature of the cultural landscape in The Peak is the network of stone walls which extend over almost the entire surface. It is perhaps the most distinctive as well as the most widespread of man's contributions to the upland scene, but however appropriate this system of fencing may appear from the aesthetic standpoint, its origin is solely functional.

In relation to upland farming, whether in The Peak or elsewhere, stone walls have a threefold purpose, firstly to control the movement of livestock and protect the arable land, secondly to assist in organising the grazing and lastly to provide shelter for the animals. As a method of fencing, dry-walling is at once the most practicable and the least expensive. On the higher exposed parts hedgerows are difficult to establish; they suffer damage from storms and in any case require far more attention than walls. Even modern post-and-wire fencing has proved more expensive and does not last so long, whereas the great merit of the stone wall is that it will stand for generations and may even persist for fifty years without need of repair. The material, whether gritstone or limestone, is everywhere at hand, and wherever walls have been erected the stone has always come from the immediate vicinity. Loose stones picked from the fields are mainly used for filling because of their worn edges and irregular shapes. Freshly cut stone enables roughly rectangular blocks to be used and makes the walls compact. The composition of the walls, as with the buildings, reflects the local rock type. The grey-white limestone walls contrast sharply with the dark brownish tinge of those made from the shaly grits and the even darker brown of those built of rough, partly rounded blocks from the massive grits. Occasional dark 'toadstones' found in limestone walls betray the occurrence of nearby igneous rocks.

Thus in the Peak District dry-walling constitutes almost the only method of fencing above 900 feet. On lower ground it is still characteristic but below 750 feet hedgerows are not uncommon. On the highest ground, especially above 1,500 feet, walling is less frequently seen because much of the moorland at this elevation remains unenclosed.

Most of the existing walls were built as a result of the Enclosure Movement not earlier than two centuries ago. Though there were a few previous instances, the movement did not begin effectively in The Peak until 1760, after which it progressed steadily until 1824. Some forms of enclosure, however, took place far earlier. In Tudor times flock-masters raising sheep for the wool trade built walls to limit the sheep runs, while even earlier the monastic granges enclosed at least part of their land by stone walls, thus separating the corn land from open grazing. It is doubtful, however, if any of these walls remain today.

In building a stone wall one or two foundation courses are laid below ground level and upon these the wall is erected, with a rubble filling, to a height of about 4 feet 6 inches. Farm boundary walls, as distinct from those dividing the individual fields, are built rather higher, to 5 feet 3 inches or 5 feet 6 inches. At the base the breadth is about 27 inches but is a little narronwer at the top. To give solidity, 'through' stones traversing the full breadth of the wall are placed at more or less regular intervals, while flatter stones are laidedgewise along the top to form a coping. An interesting detail is the occasional omission of a stone near the ground, giving an opening known as a 'creephole' through which sheep may pass from one field to the next. To be serviceable the creep hole should be large enough to permit the passage of a ram or a ewe when carrying her lamb, but in very old walls they are now too small because sheep have been bred to an appreciably larger size than in the past.

Dry-walling, though not entirely a dying craft, requires considerable skill and expert builders are fewer than formerly. Competitions are held periodicallyand spectacular results are still achieved in constructing a length of wall in a given time. A moment's reflection will show what an immense amount of labour has been involved in establishing the vast pattern of walls which form such a characteristic element in The Peak landscape. In the limestone country an average-size farm of 75-80 acres with typical five-acre fields requires three miles of walling, and a single square mile of such land involves an aggregate length of 24 miles.

Very successful dry-walling of recent date can be seen in the neighbourhood of the Hope Valley cement works. Agreements relating to land for future quarrying in connection with the industry have resulted in the modification of local farms and the need to fix new boundaries. The latter have been established in the form of stone walls in the traditional manner and are splendid examples of the art.

To the careful observer the arrangement of field walls varies from one part of The Peak to another. In fact to some extent the walls provide an index of agricultural development. This is best illustrated by reference to three distinct cases. In the first case the pattern of relatively small and slightly irregular, squarish fields which are characteristic of the limestone plateau and elsewhere, reflects the change from the early open-field agriculture to the more modern holdings promoted by the enclosure movement, broadly during the period 1760-1830. Here and there, however, groups of rather narrow fields can be seen separated by walls which, although running parallel with each other, are slightly curved in the form of a reversed S. Now Dr S. R. Eyre has shown that such a curve is clear evidence of open-field strip cultivation. In such cases, therefore, it may be assumed that early and rather piecemeal enclosure led to such parcels of land being fenced (or walled) just as they stood in the original strips. The walls have thus served to fossilise them as fragments of the ancient system. Good examples of this effect of early enclosure may be seen along the east side of the Derwent Valley at Beeley, Baslow and Hathersage. Thirdly, on the higher moorland, in contrast to the other two cases, long straight gritstone walls divide large rectangular fields with almost geometrical regularity. Related to this pattern are the straight lengths of drive and track which intersect at right-angles. This aspect of the moorland landscape is the product of systematic enclosure under the Commissioners administering the later Enclosure Acts of the nineteenth century. Truly there are sermons in stones.

REFERENCES

Couzens, F. C. The Personality of Settlements to North Derbyshire. *Derbyshire Countryside* (July, 1938)

Couzens, F. C. Distribution of Population in the Mid-Derwent Valley since the Industrial Revolution. *Geography: 26* (1941)

Tate, W. E. Enclosure Acts and Awards relating to Derbyshire. *Jour. Derbys. Arch. and Nat. Hist. Soc.: LXV* (1944)

Thompson, F. *A History of Chatsworth.* Country Life (1949)

Powell, A. G. The 1951 Census: an Analysis of Population Changes in Derbyshire. *East Midland Geographer: 2* (1954)

Maxwell, I. S. Section 7, The Age of Settlement, from *Sheffield and its Region.* British Association for the Advancement of Science (1956)

Eyre, S. R. The Upward Limit of Enclosure on the East Moor of North Derbyshire. *Trans. Inst. of Brit. Geographers* (1957)

Cameron, K. *The Place-Names of Derbyshire.* English Place-Name Society, vols XXVII-XXIX (1958)

Raistrick, A. *Pennine Walls.* Dalesman Publishing Co. New ed. (1961)

10. Peakland Towns and Routes

He likes the country, but in truth
Most likes it when he studies it in town.
W. Cowper: Retirement

The shape and extent of the Peak National Park are largely a reflection of the structure and drainage system of the southern Pennines. This is partly owing to the doming of the Carboniferous Limestone and partly to the predominantly southward trend of the chief rivers. The result is that the marginal lowlands to the east and west, with their industrial areas, lie relatively far apart. The towns of the one are completely separated from those of the other by the intervening uplands, whereas farther north the trans-Pennine valleys permit of considerable urban and industrial penetration from both sides. By a judicious placing of the boundary the Park thus provides a maximum extent of open country with a minimum amount of urban development. In fact within its compass the only urban centre of any size is the small town of Bakewell with 3,600 inhabitants.

Immediately outside, however, at a number of points, small or medium-size towns lie near to or even impinge closely upon the boundary, as does the large city of Sheffield. Such towns are Matlock, Wirksworth, Ashbourne, Leek, Macclesfield, Buxton, Chapel-en-le-Frith, Whaley Bridge, New Mills, Glossop, Holmfirth and Penistone. Only a little farther away are larger centres such as Chesterfield to the east, Stockport and Oldham to the north-west, and Huddersfield to the north. Finally, at varying distances, come the major centres of Manchester, Derby and Nottingham. Sheffield, the other major centre, has spread so far to the west that part of the municipal boundary actually lies within the Park, while for many years past a number of its people have taken advantage of good road and rail access to become resident at Grindleford and Padley in the Derwent Valley.

The pattern formed by the encircling groups of towns provides many approaches to the National Park. From distant parts of the country it is most easily reached through what have been termed the major centres, i.e. Manchester, Sheffield, Derby and Nottingham. Between these centres the Park is traversed by main-line railways and trunk roads, while from the large towns of West Yorkshire good roads enter the northern part of the area from Huddersfield. Of the railways, the two most important lines are those of the former Midland Railway, the one from London and Derby to Manchester via Matlock and Miller's Dale (now closed between Rowsley and Miller's Dale, including both these places), and the other from Sheffield across to Manchester via the Dore and Totley tunnel and Edale. The most frequented road routes are those connecting Derby with Manchester via Buxton (A6), and Sheffield with Manchester via Glossop (A57) or via Longdendale (A628).

Smaller towns such as Glossop, Macclesfield and Leek on the west side and Chesterfield on the east, all of which are situated so close to the Park that their outskirts now reach almost to the boundary, serve as further points of entry. From each of these a good road leads into the interior through impressive scenery. From Glossop the road to Sheffield (A57) mounts over the Snake Pass into the moorland country of the upper Derwent and its tributaries. The road from Macclesfield winds its way up to the West Moor around the flank of Macclesfield Forest and on to the well-known Cat and Fiddle Inn before descending to Buxton. That from Leek to Buxton passes close to the Roaches and thence along the flank of Axe Edge, giving fine views over the limestone country to the east. On the other side, Chesterfield, the largest industrial centre of the Derbyshire coalfield, is certainly a gateway to The Peak, for a few miles of well-graded road over the East Moor give quick access from the town to Baslow, Chatsworth and Bakewell.

Of the towns to be mentioned in a little more detail, Buxton and Matlock come first as being closely connected with the activities of the National Park and with the tourist traffic which the Park attracts. Then come Ashbourne, Wirksworth and Holmfirth, three of the smaller marginal towns, followed by Bakewell, the chief town within the Park itself. Lastly, reference is made to Castleton, Tideswell and Hartington, which are examples of still smaller centres possessing some particular interest.

BUXTON

Of all the towns intimately related to the National Park, Buxton holds a curiously anomalous position. For although it is the leading resort of the Peak District, it is not, for the reason already given in chapter 1, included in the Park. The exclusion of the Buxton district causes several other features of interest to fall outside the Park. Such are Poole's Cavern, one of the former wonders of The Peak, now closed to the public, the viewpoint on Grin Low (Solomon's Temple), and the Bull Ring at Dove Holes. The hill fort on Combs Moss is just inside the Park. In effect, however, except for national park jurisdiction and related planning control, Buxton and its neighbourhood are inseparable from the Peak Park. As a focus of communications and as a centre for the accommodation and entertainment of visitors, the town has no equal.

Just as its thermal springs were the most important among those in the region to be exploited by the Romans, so also Buxton was the first of the modern spas in The Peak. Even in Tudor times the baths were frequented by invalids whose names and symptoms were regularly recorded, and in 1572 a certain Dr Jones of Derby produced a treatise on the curative properties of the springs. Despite alterations, the Old Hall, now a hotel, stands as a witness of this period of Buxton's history. From the middle of the sixteenth century the town has been closely dependent upon the patronage of the Dukes of Devonshire. With the completion of the famous Crescent in 1784, built by the fifth Duke to the design of John Carr of York, new facilities and a new elegance were given to the spa. The Crescent, with its classical style, was designed to rival James Wood's Crescent at Bath and its erection, actually on the site of the Roman baths, marked the real beginning of the modern resort which soon grew rapidly as a fashionable watering-place.

The thermal springs occurring in the limestone issue at a temperature of nearly 82° F. (those at Bath being 119° F.) and are chiefly used by sufferers from rheumatic complaints, gout and arthritis. In addition, water from the chalybeate spring which occurs in the shales at the foot of Corbar Hill was at one time conveyed by pipe to the Pump Room and was valued as an iron tonic for cases of anaemia. The Pump Room is now used as an information centre.

While the thermal springs, including the historic St Anne's Well, which gave rise to the spa, occur along the valley floor close to the Wye, the old settlement of Buxton, including the market-place, developed on higher ground to the south. Although they have grown into one, these separate elements of the town, Higher and Lower Buxton, are sufficiently different in character and appearance to remain distinguishable. In the upper town the small seventeenth-century chapel of St Anne recalls an earlier link between the two, for it was built to meet the spiritual needs of those seeking benefit from the waters below before the spa had fully developed. Also in the upper town, attached to the public library, is a museum of natural history and archaeology relating to The Peak. As the highest market town in the country, Buxton has a meteorological station placed at an altitude of 1,007 feet, the records of which, allowing for a change in the site of the station in 1923, date back to 1868 and are of particular interest in the study of upland climate.

With its present population of 20,000 Buxton is a significant urban centre. It is the focus of shopping, entertainment and other services for a considerable residential population and for the outlying communities concerned in the quarrying and lime-extracting industry. Though less significant as a spa than it was formerly, the town is still important as a holiday centre. Since the bulk of its visitors spend much of their time seeking the attractions of The Peak, Buxton is as much a part of the Park as it is a gateway into it.

MATLOCK

Matlock, on the south-east border, occupies a somewhat similar position to that of Buxton and was excluded from the Park on similar grounds. The collective name of The Matlocks, by which it used to be known, gave a hint of the succession of places called Matlock Bank, Matlock Town, Matlock Bridge, Matlock Dale and Matlock Bath respectively, strung along the Derwent Valley, which have not yet fully coalesced to form a continuous town. So far as modern Matlock is concerned, the development of two of these centres, Matlock Bank on the steep eastern slope rising from the Derwent bridge and Matlock Bath about two miles further down the valley, largely accounts for the growth of the town. Both owe their origin to hydropathic establishments which gained them a reputation as spas.

The earliest settlement, however, known as Metesford, was located on a spur of limestone which forms a northward continuation of the High Tor cliffs. Metesford occupied a vantage point from which it controlled the crossing of the Derwent by one of the ancient roads from Chesterfield to Buxton. It is not known for certain when the change of name took place but it was probably in the thirteenth century. Something of this part of old Matlock, now known as Matlock Town, survives in the cluster of buildings around the parish church of St Giles. A small village green, an eighteenth-century house on the site of the original manor, an inn or two and groups of closely-built gritstone houses provide a reminder of the days when the place was primarily a centre of quarrying and lead-mining. After the ford was replaced by a bridge another nucleus of settlement was formed, eventually becoming known as Matlock Bridge.

As a spa Matlock owes much to two early industrialists, one in the eighteenth century and the other in the nineteenth. It appears that the special qualities of the tepid springs issuing from the limestone in the gorge south of the High Tor, were not recognised until about 1690, when they were noticed by lead-miners. A bath house erected in 1698 was successful in drawing visitors, and in this way Matlock Bath was founded. But the new spa suffered for a long time from a major disadvantage in not having adequate road access along the Derwent Valley from Derby and the south. This drawback was eventually overcome through the initiative of Sir Richard Arkwright, who was mainly responsible for the construction of the road between Belper and Cromford which was later extended to Matlock by cutting through the limestone spur at Scarthin Nick. This enabled the road to follow the twists and turns of the river, passing through Matlock Bath on the way to the Bridge. The road became a public highway in 1818 and ensured the growth of the spa. Later on, with the building of the Midland Railway from Derby and Ambergate along the Derwent Valley to Rowsley in 1849, the future of Matlock as a resort was assured. Now another industrialist was to play his part. John Smedley, who owned hosiery mills at Lea Bridge, near Cromford, became interested in hydropathy and directed his attention to Matlock Bank. Here in 1853 he started building the huge hydro which still bears his name. Smedley's success soon induced others to build similar establishments for the treatment of rheumatic and allied complaints. Smedley himself erected Riber Castle as a private residence in 1860, a building at once eccentric and pretentious, which even more than the hydro, dominates the town. The west wing of the hydro, like Riber, is battlemented in accordance with Smedley's romantic notion of medieval splendour, but the east wing, built after his death, is in neo-classical style. The hydro has recently been closed and is now used for the administrative offices of the Derbyshire County Council which were formerly in Derby.

b

9. FUNGI

a. Lawyer's Wig (*Coprinus comatus*). *b*. *Amanita rubescens*. *c*. *Armillaria mellea*. *d*. Orange-peel Elf Cup (*Peziza aurantia*)

c

d

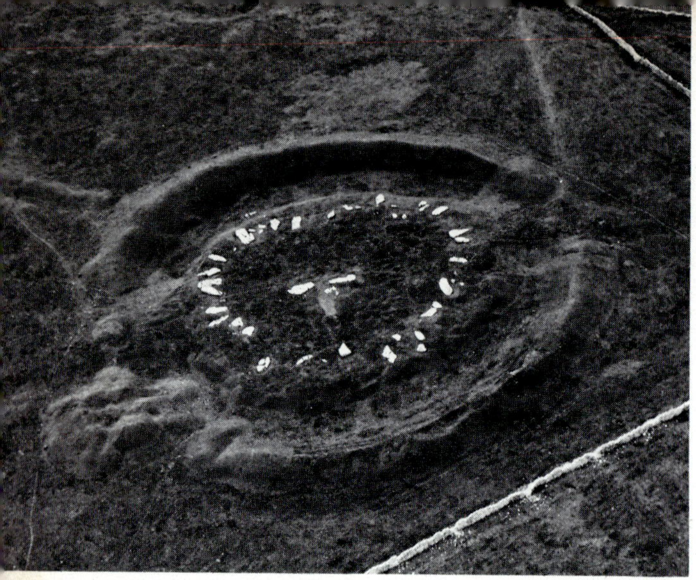

10. *Above*. Arbor Low Stone Circle. *Below*. Five Wells, a Neolithic burial chamber near Chelmorton

11. *Above*. Chelmorton. Fields bounded by stone walls indicate early and late enclosure. The long narrow fields show early enclosure based on old cultivation strips; the longer squarish fields were enclosed later. Trees, chiefly for windbreaks, were planted still later. *Below*. Hartington Hall, a typical stone-built manor house of the Tudor-Jacobean period

12. *Above*. Cottages at Chelmorton, characteristic of the limestone country: the house and the dry walling are of limestone but the roofing flags, door and window lintels, and gateposts are of gritstone. *Below*. Modern cottages built in traditional stone, Bakewell. The cottages were erected in 1968

13. *Above*. Bakewell. The church stands on the slope at the extreme left, while Castle Hill on the opposite side of the valley overlooks the town. *Below*. The church of St John the Baptist, Tideswell, the 'Cathedral of the Peak'

14. Castleton, at the southern edge of the Hope Valley (Edale Shales); in the foreground is Cave Dale (Carboniferous Limestone), while in the background is Lose Hill (Millstone Grit) reaching to 1,563 ft

15. Deep Rake and High Rake, old lead workings near Calver. Deep Rake with the narrow tree belt, mostly of beech and sycamore, continues in the distance as High Rake. The former is being re-worked for fluorspar

16. *Above*. Unfinished millstones, Burbage Edge. *Below*. A new use for millstones, marking the entrance to the National Park, Chesterfield-Baslow road

An important difference between the two spas should be noted. At Matlock Bath natural springs in the limestone having a temperature of about 68° F. are utilised, but at Matlock Bank, Smedley and the other hydro-owners drew their supply of soft water from the junction of the gritstone and shales on the moor above the town. Thus Matlock Bank became a centre for hydropathy, using the water mainly for external application in various forms of treatment, whereas Matlock Bath exploited the medicinal properties of its thermal springs. The contrast between the two centres is heightened by the fact that at Matlock Bath all the phenomena of limestone scenery are present: cliffs, caves, fissures, underground streams and petrifying wells. These have all been exploited commercially and have furnished the basis of present-day popular entertainment hardly in keeping with the once genteel relaxations at Matlock Bank.

During the past thirty years, Matlock, even more than Buxton, has become a popular resort and week-end excursion centre, catering for less refined tastes than in its heyday. Its glories as a spa have departed; buildings housing the baths now provide less sedate forms of recreation and many of the hydros have been converted to other uses. Industrial activities such as textiles and engineering have expanded, while limestone quarrying has produced unsightly scars. On the other hand, much has been added to the town's amenities and the change in the basis of its economy has brought rapid growth. The town is a growing centre for education with a College of Education and several private boarding schools. The Urban District, which now extends along the Derwent from Matlock Bath to Darley Dale, has a population of over 20,000.

ASHBOURNE

Ashbourne, a mile or so beyond the southern extremity of the Park, is commonly regarded as one of the Peak District towns. It stands in the valley of the Henmore Brook a little above the confluence of that stream with the Dove. At Ashbourne the Henmore Brook leaves the Carboniferous Limestone and enters the Trias area of the Midlands, cutting deeply into the Bunter Sandstone where the town is situated. After crossing the narrow outcrop of the Sandstone the stream passes on to the Red Marl formation on reaching the Dove. From the physical standpoint, therefore, Ashbourne's position is marginal to the Peak District, and this is reflected by the considerable use of red brick for building, the Marls providing the clay, although many of the oldest structures are of stone. The town nevertheless marks one of the approaches to The Peak, for it gives access to Dovedale and the Manifold Valley from the Buxton road. It is of some interest, too, on its own account. Outstanding among several historic buildings is the beautiful church, the tower of which is surmounted by a tall, slender and particularly graceful spire often referred to as the 'Pride of the Peak'.

Ashbourne has had a long history as a market-town placed between the hill country and the lowland. For almost two centuries it has handled the products of dairy-farming and at various times during this period cheese fairs and butter markets have been held in addition to the general market. This activity is reflected today in a large dairy factory which, besides having a greater daily intake than any other centre in The Peak, has by far the greatest processing capacity. Since the war, however, much of this has been devoted to the production of Nescafé. A few other small-scale industries and various service functions make Ashbourne, with its population of 5,600, a typical country town.

WIRKSWORTH

Wirksworth, like Ashbourne, also lies outside the National Park. It stands at the head of the Ecclesbourne valley in the limestone country and was for a long period the principal centre of the lead-mining industry. It remains a small market-town of about 5,000 people, but is mainly dependent upon large-scale quarrying in the neighbourhood. It is still the seat of the Barmote Court, which deals with all claims relating to lead-mining. The present Moot Hall, which is also used as a nonconformist chapel, was built in 1812. It took the place of a hall erected in the market-place in 1773, which in turn had replaced a much earlier one. Wirksworth was served by the High Peak Railway and later connected with Derby by a branch line from Duffield. The west side of the town is dominated by huge quarries. Lime kilns, crushing and asphalt plants add to the industrial atmosphere. At Middleton, an adit mine produces 250,000 tons annually of exceptionally pure limestone for use in glass-making, while an overhead conveyor runs from the largest quarry to the sidings of the Duffield Railway. Road transport is also much used, especially for asphalt and limestone for construction. The earlier textile-working survives, the town being an important centre for tape production. Wirksworth is far from stagnating, and, with new housing developments, is extending.

HOLMFIRTH

Brief mention should be made of Holmfirth, the small West Riding town which lies close to the northern boundary of the Park. It stands in the narrow valley of the River Holme, a tributary of the Colne, and is hemmed in by moorland on either side. Like Huddersfield (only six miles away) and other towns in this part of Yorkshire, Holmfirth is a centre of the woollen industry and now has a population of over 18,000. It was formerly a market-town, while many old quarry sites in the local Millstone Grit testify to another activity which has declined in importance. The road from Huddersfield into The Peak passes through Holmfirth, giving access to the northernmost moors in the National Park. While the people of Leeds and Bradford are drawn to the Yorkshire Dales for open-air recreation, the people of Huddersfield find the Peak District more accessible and they reach it immediately on leaving the outskirts of Holmfirth. On fine days many motorists, hikers and picnic parties make for Heyden Moor and Holme Moss, the BBC television station on Holme Moss being an object of considerable interest.

BAKEWELL

Bakewell, in the Wye Valley, with a population of only 3,600, is the largest urban centre within the Park and is from all points of view one of the most attractive of the larger settlements. The town is situated at a bridge-point where the Wye, on emerging from its narrow course through the limestone plateau, develops a relatively wide flood-plain on approaching its confluence with the Derwent (Fig. 10). A short distance upstream from the town bridge, itself an exceptionally fine structure, is an ancient and graceful pack-horse bridge, one of the many still remaining in Derbyshire.

Although it has been an important market centre for many centuries, Bakewell is still active in this respect and serves a large country district which spreads over much of the limestone territory as well as part of the gritstone area to the east. For all this district the town is the chief retail and service centre, besides being a focus of small industries. Old writers often referred to Bakewell as the metropolis of The Peak.

The pleasantness of the surroundings, whether farmland, woodland or water-meadow, is largely due to the high proportion

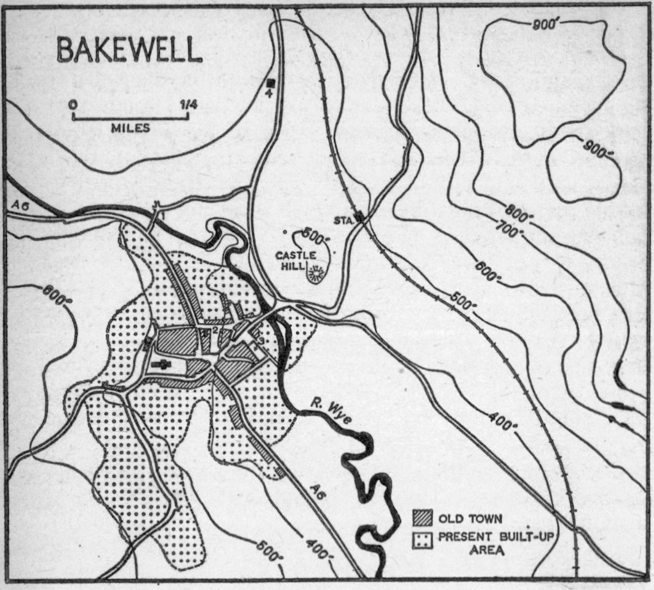

Fig. 10 Site and plan of Bakewell
1. old pack-horse bridge. 2. old bath-house. 3. cattle market. 4. Peak Park Planning Board Office. (Based by permission on the Ordnance Survey. Crown Copyright reserved)

of estate land in the neighbourhood, belonging to the Dukes of Devonshire and Rutland, which has been maintained in accordance with the best traditions of the eighteenth-century English landscape. Another of the attractions of Bakewell is the continued use of local stone for building. Here, even in the post-war housing, the local authority has set an admirable example to other places in maintaining dignity and harmony in the recent development of the town.

As the name suggests, Bakewell originated on the site of natural springs in the limestone. These included a chalybeate spring which in all likelihood was known to the Romans but, apart from that period, no attempt appears to have been made to exploit it until 1697, and even this was a short-lived venture. The temperature of the water, at 60° F., was much below that of the Buxton springs and even lower than those at Matlock. About 1818, however, following the completion of the road from Derby through Matlock and Bakewell to Buxton, new baths were erected by the fifth Duke of Rutland. By this time, however, Buxton was already a fashionable watering-place and Matlock was also growing in reputation as a spa, so that Bakewell was never able to achieve more than modest success. The Bath House can still be seen, while the Gardens form another of the attractive features of the town.

Besides serving as a base for visitors who go shooting on the moors, Bakewell has long been a noted resort for anglers, both the Wye and its tributary, the Lathkill, being well stocked with trout and grayling. Its importance as a tourist centre is steadily growing, mainly for ramblers and coach excursions, though it is also visited by holidaymakers. Having good road access to all parts of The Peak, Bakewell has proved to be a suitable centre for the National Park administration and for the seat of the Park Planning Board. In the old Market Hall, with the interior redecorated, is the Information and Exhibition Centre.

SMALLER CENTRES

Castleton, situated in the Hope Valley just beneath the northern edge of the limestone country, lies in the midst of scenery diversified by the Edale Shales which form Mam Tor, the grit-shale ridge from which rise Back Tor and Lose Hill and by the impressive gorge of The Winnats cut in the limestone. Above all, Castleton is noted for its limestone caves, several of which rank among the largest and finest in The Peak. Caves and scenery have long made it attractive to visitors, especially since the railway came into the Hope Valley and still more so, of course, during the period of motor transport. The ruins of William Peveril's castle, partially defended by a deep cleft in the limestone, dominate the village. On holiday occasions it is as thronged with people as are the nearby wonders of the subterranean world. Castleton today is chiefly a centre for day and week-end visitors and a calling-point for coach excursions. Some indication of its popularity is shown by the fact that in recent years the castle has been visited on an average by 12,000 people annually, the Peak Cavern by 17,000 and the Blue John Mine by 35,000. To the naturalist, Castleton offers a further point of quite different interest, for in the churchyard is the grave of Elias Hall, a well-known Derbyshire naturalist. Hall was a native of Castleton and died in 1853. Though self-taught, he devoted a long and energetic life to naturalist studies and attained a reputation as a geologist. He contributed to our knowledge of the stratigraphy of The Peak, made instructive models of both the Peak District and the Lake District, and produced a geological and mineralogical map of the 'Midland Coal Field of England', embracing the Yorkshire and Derbyshire and the South Lancashire fields. Hall has been called 'the father of Derbyshire geology', but an earlier geologist, White Watson who died in 1835, surely has a stronger claim to this title. Watson was associated with the famous black marble industry at Ashford near Bakewell. In 1811 he published his *Delineation of the Strata of Derbyshire* which, bearing in mind that most of it was compiled before the turn of the century, was the first comprehensive work of its kind relating to The Peak.

Tideswell affords an interesting example of the persistence of an old town in the face of adverse modern conditions. It stands in the heart of the limestone tract north of the Wye, seven miles from Buxton and a similar distance from Bakewell and is sited near the head of Tideswell Dale. The old town, long and narrow in form, lies in the dale below the level of the surrounding plateau. Its origin as a settlement depended upon permanent springs, and for centuries Tideswell served as a market centre for this part of Derbyshire and derived added wealth from lead-mining and stone-quarrying. Its fourteenth-century church in a finely ornate version of the Decorated style, is known, on account of its great size and architectural distinction, as 'The Cathedral of the Peak'. The lead trade has long ceased, quarrying has largely gravitated towards Buxton, the textile works of yesterday have closed, save for Litton Mill in Miller's Dale, while main road traffic avoids the town. Yet Tideswell, though only of village size, with 1,800 inhabitants, still has the appearance and the features of a small urban centre. A few new enterprises prompted by war-time considerations, of which the chief was the making of plastics, stemmed the economic decline but have since been removed.

Though now only a village, Hartington once had greater importance as can be deduced from many of its older buildings and its market-square. Its position overlooking the Dove Valley at the head of Beresford Dale commanded a crossing of the Dove into Staffordshire. At different periods Hartington was associated with important mineral workings, lead from veins occurring to the north, copper from the famous Ecton Mine overlooking the Manifold Valley and ironstone from Hard Dale to the east. The Dove, made famous among anglers by Isaak Walton and Charles Cotton in the seventeenth century, and the beauty of the dales through which it flows combine to make Hartington a tourist centre. Apart from this the village is mainly an agricultural settlement and the small factory producing Stilton cheese emphasises the predominance of milk production in the area.

COMMUNICATIONS

Until the mid-eighteenth century transport throughout the Peak District was largely restricted to movement by pack-horse. Many of the tracks used for that purpose had survived from ancient times and the pack-horse bridges over the streams which remain today are a reminder of this prolonged phase of communications. Built of local stone, many of these bridges, like that at Bakewell and the sheep-wash bridge at Ashford-in-the-Water, are structures of great beauty, combining strength and elegance. Others which might be noted are at Milldale (River Dove), Youlgreave (Bradford Dale), Conksbury (Lathkill Dale) and the Fillyford Bridge over the Wye below Haddon Hall. Similar relics of earlier movement, which are still in use for travellers on foot, are the small bridle bridges and slab bridges over the moorland torrents and the occasional stepping-stones like those set across the Dove at the foot of Thorpe Cloud. Examples of the bridle bridge are to be found below Jacob's Ladder, Kinderscout, over the Burbage Brook near Longshaw, and across the upper Dove at Washgate, while typical slab bridges span the Bar Brook (near Baslow) and the Bradford river in Bradford Dale. The bridges are not the only reminders of the pack-horse days, for here and there one can find place-names which originated with that form of traffic. Such is Jaggers Clough near Hope, the 'jagger' referring to the driver or leader of a pack-horse convoy.

THE TURNPIKES

The roads of the Turnpike Trusts brought a substantial improvement in communications and greatly reduced the isolation of the region; and a little later the transport of characteristic products such as lead, stone and lime was further assisted by canals. Incidentally the famous canal engineer, James Brindley, was born at Tunstead near Wormhill, and a memorial to him erected over a spring may be seen in Wormhill village, though more recently a plaque has been added to the birthplace. Not far away, near Leek, is a different kind of memorial. This is Rudyard Lake, a reservoir constructed some years after the engineer's death in 1772, by impounding a small tributary of the River Churnet. It was built to supply water to branch canals of the Trent-Mersey Canal with which Brindley himself was so much concerned. The Cromford Canal, completed soon after 1790, provided an outlet from the Derwent Valley to the Trent, while the Peak Forest Canal (1800), which served the mining and quarrying interests of the Buxton area afforded a route to Manchester and the Mersey. A remarkable achievement was the linking of these two canals in 1830 by the Cromford and High Peak Railway, the first line to be built in The Peak. Later it connected with mainline railways and no longer with waterways; but this line no longer operates. We shall return to it a little later.

As the progress of the Industrial Revolution brought new forms of economic activity into the region, so the need for improved communications grew. In this the turnpikes played a significant part. When Arkwright established his first cotton mill at Cromford in 1771 there was still no adequate road along the Derwent Valley between Belper and Matlock. When other textile mills followed, in both the upper Derwent and the Wye Valleys, access to them was still difficult. A major improvement was accomplished, therefore, between 1811 and 1817, by the conversion into a public highway of a private road along the Derwent from Milford and Belper (where the mills belonging to Arkwright's partner, Jedediah Strutt, were situated) to Cromford. The road, still following the river, was extended beyond Cromford to Matlock and Darley Dale. At Cromford it pierced the limestone spur of Scarthin Nick by a narrow cutting which, as many motorists know, seriously constricted the flow of present-day traffic until it was widened in 1963. This road was later continued along the Wye Valley to Buxton and ultimately became the trunk route between Derby and Manchester, known today as A6.

The first turnpikes to be opened within the Peak District, however, appeared in the first half of the previous century. Soon after 1720 roads were opened from Manchester to Buxton and Chapel-en-le-Frith, while the earliest to enter the region from the south was that from Derby to Brassington. Why, it may well be asked, should the latter terminate at Brassington? The reason becomes a little more apparent if we note that for a considerable time the roads from Manchester went no further than Buxton and Chapel. In all three cases the turnpikes entering The Peak from outside terminated as soon as they reached the limestone upland. This was uundoubtedly because the existing roads over the dry limestone surface were in a comparatively satisfactory condition and it was assumed that movement farther into the interior would not be difficult once access from the damp clays of the surrounding lowlands had been ensured. It is not surprising, however, that many travellers journeying northward should forsake the Brassington turnpike for the Ashbourne-Newhaven road when it came into use. Other turnpikes serving the Low Peak were the Matlock-Newhaven road which was a later extension of the route from Nottingham via Alfreton; the Matlock-Ashbourne road; and the Matlock-Buxton road.

Farther north, in the High Peak, turnpike roads leading across the moors from Sheffield were opened well before the close of the eighteenth century. Such were the road to Baslow, which was eventually extended to Bakewell, and that over Stanage Moor to Hathersage and Bamford, whence it was continued through the Hope Valley to Chapel-en-le-Frith. Here it should be mentioned that the old pack-horse route from Sheffield to Bamford followed the Long Causeway, an ancient track of probable Roman origin lying north of the turnpike. A paved section of this track still forms a useful path over part of the moor. In 1818 one of the Turnpike Acts sanctioned the building of a road over the Snake Pass by which Sheffield and Manchester could be linked more directly than hitherto. This road and many others have survived to serve as important elements in the network of main highways which covers the Peak District today. Many other turnpikes, however, fell out of use, either wholly or in part. Some remain as secondary roads, others are now nothing more than quiet lanes, while a few can hardly be traced at all. Some of the present-day secondary roads show evidence of their former status as turnpikes, for it is not uncommon to see on their verges examples of old milestones which vary somewhat in pattern according to the Trust which set them up.

THE RAILWAYS

Following the vast improvement of the highways brought about by the turnpike system, the railways played an even greater part in opening up The Peak. The basis of railway development in the region, piece-meal though it was to a large extent, was the inter-connection of the three chief marginal centres of Derby, Sheffield and Manchester. The stages by which this was achieved form one of the most interesting chapters in railway history.

The first approach to The Peak was made by George Stephenson's North Midland line from Derby to Leeds in 1840. This line followed the Derwent as far as Ambergate, which is only about five miles from Wirksworth and about eight from Matlock. From Ambergate, however, it crossed to the Rother Valley and passed through Chesterfield and Rotherham but not through Sheffield. For some years the latter was served only by a branch line from Rotherham. In fact, not until the extension from Chesterfield through Dronfield and the Bradway Tunnel was completed in 1870, was Sheffield brought into direct rail communication with Derby. The railway from Derby towards Manchester quite naturally left the North Midland line at Ambergate and continued northwards along the Derwent. As in the case of the road routes, much depended on the use of the gently graded valleys of the Derwent and Wye. By 1849 the line reached Rowsley and then construction was halted. The promoters had planned to continue the line along the Derwent but the Duke of Devonshire refused to allow it to pass through Chatsworth Park. After repeated attempts to negotiate with the Duke, an alternative route along the Wye was sought, but this met with similar opposition from the Duke of Rutland at Haddon Hall. Not until the sixties was an agreement madee to tunnel beneath the grounds of Haddon and to skirt the town of Bakewell on the east side. These and other conditions kept the track well above the level of the Wye and indeed entirely away from it in some places, making its construction correspondingly more difficult and costly. It is amusing to recall that after consent had been obtained for building the line, the two dukes were again involved in a controversy, this time between themselves. One wanted the station at Bakewell and the other insisted that it should be at Hassop. The result was that both were built! In 1867 this railway at last reached Manchester via Miller's Dale and Chinley. A few years earlier the line from Rowsley through Miller's Dale had been built to terminate at Buxton.

Thus both Sheffield and Manchester became linked with Derby and both railways provided express routes to London. Much earlier, Sheffield and Manchester were connected with each other by the line through the Woodhead tunnel, opened in 1846. A second connection was made in 1893-4 by means of the Dore and Totley tunnel (6,230 yards), which gave access from Sheffield to Hope and Edale. From Edale the Cowburn tunnel (3,702 yards) enabled the line to reach Chinley, where it joined the main route from Derby to Manchester. This did much to reduce the isolation of Edale, for until the railway came this valley was without regular communications. Even Castleton had a coach service to Sheffield three times a week, which also served Hope. Neither was the road from Chapel-en-le-Frith into Edale used at all frequently, for over Mam Tor it presented problems of slipping and subsidence just as it does today.

The other railways developed in The Peak were not 'through' lines but were designed to serve particular districts. The Ashbourne-Buxton line closely followed the track of the turnpike connecting those towns. Largely for that reason modern road transport has caused traffic to dwindle to such an extent that in 1954 passenger services on the railway ceased. It is an interesting reflection of conditions in The Peak, however, that the railway is under an obligation to run trains between Ashbourne and Buxton in the event of the road becoming impassable because of ice or snow. An emergency service was in fact put into operation for several days during severe weather in late February, 1955. The line is now closed and the metals removed to make way for the Tissington Trail. The branch railway to Wirksworth, leaving the main Derby-Manchester line at Duffield, provided an approach to the limestone upland comparable with the old turnpike to Brassington. Apart from passenger traffic the line was useful for the carriage of goods between Derby and Wirksworth and it is still in operation for this purpose, although regular passenger services were withdrawn in 1947.

The Leek and Manifold Light Railway, completed in 1904, was designed to provide an outlet for minerals and farm produce from the upland district lying west of Hartington, as well as to carry passengers. Its terminus was at Hulme End on the Manifold less than two miles from Hartington, and the track followed the winding valleys of the Manifold and Hamps southwards as far as Waterhouses, where a standard gauge branch of the former North Staffordshire Railway completed the connection with Leek. In particular the promoters hoped that the line would convey increasing quantities of milk from the farms and villages, and at the same time stimulate a revival of lead-mining in the neighbourhood of Ecton Hill, where the historic copper mine and several lead workings had proved so profitable in the past. They were doomed to disappointment for the Manifold Railway, though it struggled for thirty years, was never successful. Even the route selected for it held dubious prospects. The narrow gauge portion was about eight miles long and the branch line to Leek nearly ten miles, yet the distance by road from Hulme End to Leek was little more than eight miles. The verdict of a labourer, given at the time of the opening, was indeed prophetic: 'It's a grand bit of line but they wunna mak a go on it, for it starts from nowhere and finishes up at same place.'

Neither the mines nor the farms contributed much freight apart from milk, while as regards passenger traffic, the line being for the most part deep in the valley, was inconvenient for village folk living on the plateau. In summer, however, the Manifold Railway attracted numbers of sightseers, for the journey lay through delightfully verdant country interspersed with glimpses of the limestone crags along the valley. Nevertheless, despite absorption into the L.M.S. system in 1921, losses continued to mount and soon afterwards nearly all the milk traffic was diverted to road transport. This was the last straw and, although the line was kept in operation for a few more years, in 1934 the Manifold Light Railway was closed. The track was acquired by the Staffordshire County Council and converted into a public footpath. This soon became a popular route among ramblers and, bearing in mind the proximity of the youth hostels at Ilam and Hartington, continues to be a valuable amenity.

THE HIGH PEAK RAILWAY

Approaching Cromford from the south along the A6 road, the trace of a railway climbing a steep slope on the left-hand side can be seen. This is the beginning of the famous Cromford and High Peak Railway which started from below the road on the opposite side. As already mentioned, this railway, which was in use for over 130 years, was primarily built to link Cromford Canal in the Derwent Valley with the Peak Forest Canal at Whaley Bridge on the western side of the uplands. By this means the waterways of the Trent were connected with those of the Mersey. Actually it was the only railway ever built in this country to link two canals. It was one of the few alternatives to a proposed canal along the Derwent and Wye valleys, the cost of which was prohibitive. That the railway was regarded as the equivalent of a waterway is shown by the fact that the stations along it were called wharves – Cromford Wharf, the terminus on the east side, Longcliffe Wharf serving Brassington, Friden Wharf, Parsley Hay Wharf and so on.

Considering the early date of its construction, the High Peak Railway was a considerable engineering achievement. The track was of standard gauge and had to negotiate not only the stiff gradients encountered in crossing the limestone country, but exceptionally sharp ascents to reach the upland from the valley at either end. The terminal gradients together with other steep inclines such as the Middleton and Hopton Inclines, both near Wirksworth, were dealt with by the use of stationary engines and cable-haulage. The Sheep Pasture Incline rising from Cromford Wharf presented a slope of one in eight. It is not surprising that years later tests were made on Wild Moor alongside this railway in connection with the building of the Mont Cenis tunnel through the Alps. While the inclines were worked by stationary engines from the beginning, horses were used for haulage on the level stretches and steam locomotives were not introduced until 1841.

On the plateau the winding course of the line reached its highest altitude at 1,264 feet above sea-level at Ladmanlow. This point is 990 feet above the eastern terminus at Cromford Wharf. At Parsley Hay, and for the next few miles, the track ran close to the Ashbourne-Buxton railway of later date, but on approaching Buxton the High Peak line swings away towards the Goyt Valley and Whaley Bridge. Only a grassy lane now marks this end of the railway.

Besides providing for the movement of lead and quarry products from The Peak, it was the intention of its promoters that the High Peak Railway should convey Derbyshire coal to Lancashire and Cheshire and bring imported goods from Liverpool to the Midlands. Though some success was achieved, it was short-lived, for after a few years hopes were shattered, first by the collapse of the lead trade and later by the outmoding of canals by railways. To this situation the owners responded in 1853 by extending the line for two miles below Cromford Wharf to join the Midland Railway which at that time, as we have seen, was advancing by stages along the Derwent Valley from Ambergate. Thus instead of serving as a link between waterways, the High Peak Railway became a feeder to a main line. The resulting improvement in traffic was maintained until the Derwent Valley line eventually reached Manchester. For a period of some twenty years, beginning in 1855, even passengers were carried! After its heyday in the sixties traffic rapidly declined. For many years only the eastern portion from the Derwent as far as Parsley Hay continued in use. Limestone, crushed or in blocks, and silica bricks from Friden were still conveyed to the main line, while quarry equipment and other supplies were carried in the reverse direction.

The High Peak Railway was thus a remarkable and original venture in railway development. Its long and tortuous course is repeatedly encountered by those who walk or drive over the limestone plateau. It has played its part in the economic development of the region and despite its somewhat erroneous name, it remains an interesting feature of the past communications of The Peak.

ROAD AND RAIL

The problem of transport as it affects the National Park is primarily one of passenger movement. People enter the Park both by road and railway. Travel by rail presents little difficulty and, despite the increasing numbers conveyed by bus, the railways have for many years run special trains at week-ends from Manchester, Sheffield, Derby and Nottingham for the benefit of ramblers and hikers. Movement by road, however, has grown immensely in recent years and there are signs of increasing congestion on some of the main highways. At week-ends progress on some of them is becoming difficult in places because of the heavy use made of them by three distinct types of traffic, the private car, the coach and the cyclist. Unlike the railways, the problem arises from not being able to differentiate between 'through' traffic and what might be termed National Park traffic. At present both are expanding rapidly and both must use the principal arteries, and although much of the Park traffic favours the less important roads for part of the time, travel by car and coach is becoming such a popular means of sightseeing that it now threatens to create serious congestion on holiday occasions and at week-ends. This is not the place for detailed discussion of traffic problems but reference has been made to them in order to emphasise the fact that in the matter of communications the Peak District differs from some of the other National Parks. Largely on account of its geographical position it is easily accessible from large centres of population, and at the same time is crossed by several major lines of communication which form part of the national transport system. The coming of the motorways, particularly the M1, and to a lesser degree the M5 and M6, show this clearly. The motorway from London to the vicinity of Chesterfield brings the Park within three hours driving from the capital.

REFERENCES

Adam, W. *The Gem of the Peak: or Matlock Bath and its Vicinity.* (1828, and later editions to 1857)
Williams, F. S. *The Midland Railway.* Fourth edition (1878)
Stretton, C. E. *History of the Midland Railway* (1901)
Derbyshire. *Victoria County History: vol. 2* (1905)
Tudor, T. L. Old Roads and River Crossings in Derbyshire. *Derbyshire Countryside* (April, 1938)
Heape, R. G. *Buxton under the Dukes of Devonshire.* Hale (1948)
Foster, J. The Town of Bakewell. *Report and Analysis of Survey* (Part IV). Peak District National Park Development Plan (1955)
Rimmer, A. *The Cromford and High Peak Railway.* Reprinted (1960)

11. Ploughland and Pasture

Happy the man, whose wish and care
A few paternal acres bound,
Content to breathe his native air
In his own ground.

Alexander Pope: Solitude

Let us never forget that the cultivation of the earth
is the most important labour of man.

Daniel Webster (1840): Remarks on Agriculture

The Peak District, despite its elevation and broken surface, is predominantly an agricultural region. Certainly most of the people living in The Peak depend upon farming for their livelihood. This has been the case for many centuries, in fact since prehistoric times. Among the oldest relics of agriculture still to be seen upon the landscape are the grassy banks and ledges known as lynchets which are found on hill slopes in some places. These narrow platforms, of which several often form a series running parallel to one another along the slope, are accepted as evidence of early cultivation. Precisely how they were formed and at what date is not really known. Probably the banks of soil were accumulated by downwash along the lower edge of each cultivated stretch. It is believed that lynchets, which occur frequently upon the chalk lands of southern England, may range in date from late prehistoric (e.g. Celtic) to medieval. Significantly their occurrence in The Peak as at Cross Low (Alsop-en-le-Dale), Sharplow (Tissington), Horstead (between Taddington and Priestcliff) and about half a mile to the west of Bakewell appears to be confined to the Carboniferous Limestone. As we have seen, this was the earliest part to be cleared for permanent agriculture, thus affording some parallel perhaps with the chalk areas.

Although the character of agriculture in the region has changed from time to time during the centuries, it is still true that the rock types, which form the basis of the main physiographic, soil and vegetation divisions, continue to promote variations in farming activity from one part to another. Equally important at the present time, however, is the effect exerted upon farming by the neighbouring industrial areas. By far the greatest demand is for milk, so that dairying is now the dominant form of agriculture and almost all parts of the region are to a greater or lesser degree concerned with the output of milk.

THE RISE OF DAIRYING

While the present emphasis on dairying, especially in the form of liquid milk production for the wholesale market, is common to most of The Peak, this has only been the case during the past sixty years. Dairying in general, however, is a much older activity, dating from the early nineteenth century. Since then the increasing demand for its products has gradually converted the upland farming from a sheep and stock-rearing type, through the stage of dairying for domestic cheese- and butter-making to the present-day stress on liquid milk. Until almost a hundred years ago, both arable farming and sheep-raising were of primary importance. In his *General View of the Agriculture of the County of Derby* in 1794, T. Brown estimated that 20 per cent of the limestone enclosures were under cultivation; and thereafter the arable acreage continued to increase until the early 1870's when it reached a maximum. Afterwards a decline set in and the arable acreage for Derbyshire in 1939 was only half that for 1866. The decline in sheep has been more prolonged; in 1800 there were 360,000 of the short-wool breed alone, but in 1866 the entire sheep population was 258,000 and by 1948 it was only 98,000. The increase in dairy cattle can be said to date from the enclosure movement, J. Farey recording in 1811 that 'since enclosure . . . the old limestone sheep have almost entirely given place to dairy cows or to more useful varieties of sheep.' Later in the nineteenth century, just as the lead-mining communities in the limestone area had long provided a market for cattle products, especially butter and cheese, so pastoral activity on the gritstones of the north-west and on the flanks of the East Moor began to benefit from the growth of the bordering coal-mining and manufacturing industries. These gritstone areas were the first to be concerned with liquid milk production.

In the interior, however, even before railways permitted the bulk movement of milk, the making of butter and cheese was a characteristic pursuit. Derbyshire cheese gained a considerable reputation. It is interesting to note that the first milk factories established in the country were erected in Derbyshire in 1870. These were small co-operative factories modelled on those found some years earlier in America. The first two were built at Derby and Longford in the south of the county, but others soon followed in the limestone area of The Peak, at Alstonfield, Grangemill, Holms (near Ashbourne), Hartington and elsewhere. Each co-operative factory obtains its milk from a group of farms in the neighbourhood. As regards cheese production they operated rather spasmodically and some of them, on account of the rising demand for fresh milk outside the district, were short-lived. Many continued in use as depots for the fresh milk trade, and even after this phase ended a few remained and can still be seen. The Grangemill factory made cheese until not many years ago, while that at Hartington after being derelict for a period, is again active and produces Stilton Cheese.

An enormous increase in milk production came with railway development. Soon after 1870 milk was sent regularly to London. Over eight million gallons were carried on the Derbyshire section of the Midland Railway in 1888, almost nine times the amount carried when the line was completed along the Derwent Valley in 1872. In the next decade several new lines were built; the Dore and Chinley line (1895) linking Edale and the Hope Valley with Sheffield; the Ashbourne-Buxton line (1897); and the Manifold Valley Light Railway (1904) on which milk proved the most profitable item of freight. These improvements left only one area, that around the head-waters of the Dove, Manifold and Dane, lying at a distance of more than five miles from a railway.

During the past century there has been an overall loss in the acreage under crops and grass in the Peak District and an increase in the amount of rough grazing, especially in the poorer parts. While there has been no marked increase in the size of the dairy herds, the acreage of land cropped for hay has been considerably extended and at the same time sheep and dairying districts have tended to become more specialised and consequently more clearly differentiated. As a result of the wartime policy towards self-sufficiency in cattle fodder, the arable acreage was substantially increased both for grain and root crops and, although some decline has occurred since the peak of 1943 and 1944, it is still about twice what it was in 1939. A good deal of permanent pasture, however, was lost during the war, some of it to the plough it is true, but much of it through deterioration. Today, farmers in The Peak, in common with all others receiving an assured price from the Milk Marketing Board, whatever the use to which the milk is put, derive little benefit from their geographical position. On the other hand, the industry has suffered no reversal and dairying remains the dominant form of agriculture. The high density of dairy cattle also shows that despite mediocrity of pastures and marginal climatic conditions, the productive capacity of the region is high. Certain areas are largely devoted to sheep-grazing with store cattle as a secondary interest, a type of farming closely related to physical conditions under which dairying on its own is unprofitable.

CATTLE AND THEIR PASTURES

Successful cattle grazing naturally depends on the quality of the pasture. To maintain the grassland on acid soils such as those of the gritstone country heavy applications of lime are necessary. On the limestone soil conditions are better, though on the higher parts leaching causes acidity and again lime must be applied. Also, if grazing is discontinued for a time the fields are soon colonised by the poorer grasses and later by heath. Not only lime but potash dressing is necessary to raise good crops of clover. Generally the pastures are composed of a mixture of rye-grass, fescue, cocksfoot and timothy, with some clover. On the gritstone and shales they mainly take the form of temporary leys varying in duration from 4-5 to 7-8 years, while on the limestone they are maintained for a much longer period.

In the Peak region the commonest grass is a rather poor quality *agrostis,* and efforts to establish better pasture by including rye-grass have everywhere proved a difficult task. On the higher north-western parts of the limestone plateau, in Edale and on the West Moors, mountain fescue pastures are found. In fact the only areas capable of supporting a lasting rye-grass pasture, and which incidentally yield good crops of wheat, are the alluvial lands of the Derwent and Wye.

Since the farming economy of The Peak is based on livestock and livestock products, animals rather than crops are the more satisfactory index to a real variation in agriculture. With regard to cattle, dairy animals, chiefly Friesians and more recently Shorthorn Crossbreeds, are almost everywhere more important than store cattle, even in the rough grazing districts. The areas of highest density are naturally those having a relatively large proportion under crops and grass, and these include the limestone plateau (an average of 17·5 dairy cattle per 100 acres) and the south-west shale district (27 dairy cattle per 100 acres). Moderate densities are found in the Hope and Derwent Valleys and on the northernmost shale and grit outlier. On many farms a high proportion of heifers indicates the importance of rearing dairy stores.

Cattle other than dairy herds are found in moorland districts of rough grazing as well as in the areas with plenty of land under crops and grass. Only in the valleys of the northern moors and on parts of the Abney-Eyam Moor do they equal dairy cattle in importance. Stock for rearing are usually bought from outside The Peak, chiefly from Ireland and the North of England.

SHEEP GRAZING

Sheep are distributed in marked contrast to dairy cattle for they are essentially the product of moorland grazing. The greatest density accords with the highest proportion of rough pasture, so that the northern moors (53 sheep per 100 acres, i.e. one sheep to every two acres), the Hope Valley, the Abney-Eyam Moor, the East Moor and the West Moor are the principal areas. All these districts carry permanent breeding flocks of hill sheep, chiefly Blackface and Swaledale, while the native Gritstone and Lonks remain important in the northernmost moors. The two last-named breeds are of interest because they have so long held their place among the upland flocks. For hill sheep both kinds are rather large and provide close, fine wool, yet they are particularly hardy. The Gritstone, indigenous to the hills and dales of the Millstone Grit country, is easily distinguishable by its grey fleece and black and white markings on face and legs. For Gritstones the chief breed-sales today are held at Haslingden, Lancs., in September, and at Buxton in October. Moorland grazing in general allows many farmers to keep their lambs through the winter for sale in the autumn at the annual sheep fair at Bakewell.

The predominantly dairying areas of the limestone plateau and the south-west shale district show the lowest density of sheep and the lowest proportion of sheep to dairy cattle, indicating that in general the two types of livestock farming are not complementary. In three localities on the limestone, however, sheep are important. These are around Peak Forest and Castleton, where hill breeds such as Kerry Hill and Clun are in favour;

around Longstone and Ashford; and in the south-west around Tissington and Ilam, where short-wool Oxford and Suffolk types are commonest.

In general then, dairying is the dominant type of farming on the cultivated land of all the soil types. Where rough grazing abounds and cultivation is restricted to narrow stretches in the valleys, sheep become the principal interest with a little dairying found locally on lower ground as in Edale. Over the gathering grounds of the large water supply undertakings, as in the Goyt Valley, Longdendale and the upper Derwent, the elimination of cattle in recent years to avoid pollution has further accentuated the domain of the sheep farmer.

As far as dairying itself is concerned it is useful to distinguish between those parts in which it is developed to the virtual exclusion of all other interests and those in which sheep or beef cattle form an important adjunct. In the first category there are three principal districts: (1) the south-west shale area and the southern extremity of the West Moors from which milk is sent to the Potteries and to London or to the factory depots at Reaps Moor, Glutton Bridge, Hartington and Ashbourne; (2) the limestone plateau area south of the Wye, which supplies milk to Buxton and Manchester, to Sheffield and to London; (3) the Derwent Valley between Grindleford and Baslow and the less elevated limestone area around Bakewell. In the second category are the areas in which store cattle or sheep form a valuable secondary interest. These are (1) the limestone plateau north of the Wye and the high limestone country farther to the west and south-west, (2) the Hope Valley where many farms are linked with the surrounding hill grazings and in which large numbers of sheep are wintered, (3) the West Moors, with the exception of the central portion which is devoted more fully to sheep farming.

Many farms on the higher limestone areas find their stock-carrying capacity limited by the uncertainty of water. Much of the field supply is from meres to which additional quantities must be carried in dry seasons. In a few years, however, the supply schemes of the Bakewell and Leek Rural Districts will bring piped water to most of the limestone farms. In the meantime the absence of good facilities may cause some farmers to discontinue milk production and to revert to livestock rearing.

The necessity for a large proportion of grassland explains why comparatively few arable fields are seen in The Peak, though the latter are noticeably more frequent in the limestone area. Here the development of pasture follows a well-established practice. Prior to seeding, the ground is tilled with oats for two or three years accompanied by rape, cabbage or even a few potatoes. This is organised so that only a fraction of each farm is under the plough at any one time, yet all the pasture fields are brought into cultivation approximately every fifteen years. Arable land is also fairly prominent on the extensive clay soils of the shale areas and the shale ledges between the grits along the Matlock section of the Derwent Valley. Oats and green crops and even wheat are grown on these soils while the pastures are organised as short leys of three to five years. A recent innovation is the mushroom farm, using old wartime premises, at Harpur Hill near Buxton, which gives work to many women in the district.

REFERENCES

Brown, T. *General View of the Agriculture of the County of Derbyshire.* (1794)
Farey, J. *General View of the Agriculture and Minerals of the County of Derby* (3 vols). (1811)
Rowley, J. J. *The farming of Derbyshire.* (1853)
Reports of the Land Utilisation Survey of Great Britain
Beaver, S. H. West Riding of Yorkshire. (1941)
Boon, E. P. Cheshire. (1941)
Harris, A. K. Derbyshire. (1941)
Myers, J. Staffordshire. (1945)
Fussell, G. E. Four Centuries of Farming Systems in Derbyshire, 1500-1900. *Jour. Derbys. Arch. and Nat. Hist. Soc.: LXXI* (1951)
Henderson, H. C. K. The Agricultural Geography of Derbyshire in the Early Nineteenth Century. *The East Midland Geographer: 7* (June, 1957)

12. Wealth from the Rocks

The truth of Nature lieth hid in certain deep mines and caves.
Francis Bacon: Advancement of Learning

The Peak District as a whole constitutes an important source of mineral wealth so that mining and quarrying form one of the chief economic activities. These operations moreover cause considerable modification of the natural landscape and give rise to characteristic features which are sometimes unsightly.

When the Romans came to Britain they were drawn to the wilds of the Peak District by two attractions – the warm springs and lead – the one for pleasure and the other for profit. The mention of warm springs naturally calls to mind those of New Zealand and other parts of the world. Nevertheless the two belong to quite different categories. Hot springs are associated with volcanic regions and may be regarded as belonging to the declining phases of volcanic activity. Their waters are to a large extent 'juvenile'; that is to say have been released from the interior of the earth and now appear at the surface for the first time. The warm springs of The Peak on the contrary consist of water which has descended from the surface to such depths that it has become heated and has returned to issue forth again at a higher temperature.

Though today there are no volcanoes to add to the interest of the Peak District, it was a minor volcanic region at the time when the Carboniferous Limestone was being formed. From its volcanoes lava streams issued occasionally and, flowing over the sea floor, became interbedded with the limestone. Though liquid lava is popularly called 'molten rock', it is, strictly speaking, a hot solution of many different minerals and is called 'magma'. In course of time the deeply-seated reservoir, from which the magma came, cooled and various minerals, such as quartz and felspar, crystallised out and formed solid rock-like granite. Many other minerals containing tin, copper, lead and the like remained mixed together either in solution or as vapours. This mixture poured up along cracks into the overlying rocks. These reacted chemically upon one another as they cooled and, crystallising out formed valuable ores in a more or less orderly succession. Tin ore was formed at an early stage and at a deep level, next came copper ore and at a still later stage lead ore towards the outer extremities of fissures. Thus the lead veins and lodes of the Peak District came into being. Rain, frost and rivers working ceaselessly after the Carboniferous period and later during the Tertiary era gradually removed the mighty over-burden of Millstone Grit, Coal Measures and of other rocks laid upon the region during the Mesozoic Era until at long last the very roots of the region were exposed and the valuable mineral lodes made accessible to the future miner.

LEAD AND LEAD-MINING

Lead, for which Derbyshire has so long been famous, occurs in the metallic form as galena (lead sulphide). Only very small quantities of white lead ore or cerussite (lead carbonate) have been found, though past workings from the White Rake on Tideswell Moor are recorded.

Unlike some lead ore that of Derbyshire is poor in silver. There is no real evidence that the Roman workings in the county were richer in this respect than those of later date. There is some doubt as to whether the Romans ever recovered silver as they did in Mendip, despite the findings of pigs of lead inscribed EX ARG and EX ARGENT. The Ball Eye Mine near Matlock is the only place where more than a negligible amount of silver-lead has been recorded. A very small percentage of zinc as well as traces of antimony sometimes occur with the lead.

The workable ore occurs in the Carboniferous Limestone as veins filling vertical fissures. These veins, known in Derbyshire as 'rakes', appear to follow local anticlines and have been exploited individually over considerable distances and are still traceable by the long narrow spoilbanks generally running from east to west (Fig. 11). Less frequently, lead may occur at the intersection of vertical joints in the rocks, and these concentrations are called 'pipes'. As already noted, the ore and associated minerals were deposited in fissures and joints in the limestone through heated water and vapours rising from igneous material at depth. That the mineralising agents remained in the limestone is due to the former cover of impermeable shales, lavas and other rocks preventing their escape.

Lead is accompanied by various secondary minerals called gangue minerals, chiefly fluorspar, barytes and calcite, which usually occur in a definite sequence in the vein. Very broadly fluorspar, one of the commonest of these minerals, is predominant towards the eastern end of the larger veins but diminishes in proportion westwards, while calcite and to a less extent barytes become correspondingly more abundant. In the past the gangue minerals were valueless, but they are now of considerable economic importance. Demand for them has resulted in a new phase of mineral working in the limestone area, leading to the exploitation of spoil banks where large accumulations still occur and to the re-opening of old lead mines.

Among the more important lead veins worked in the past are those of the Castleton area (Dirtlow and Moss Rakes), Stony Middleton (Dirty, White and High Rakes), Youlgreave (Long and Lady Rakes), those of the Wirksworth and Winster district, and those around Matlock.

Lead production reached its zenith in the late eighteenth century but remained fairly high until decline set in after 1850. While mines continued to close throughout Derbyshire, one venture proved exceptionally rich. This was the Mill Close Mine on the west side of Darley Dale which, after being worked in the eighteenth century, was reopened in 1861. Though the amount produced fluctuated with prevailing prices, the mine remained active for a long period; the war of 1914-18 stimulated a great increase in output which continued to rise until the early thirties, when the quantity of ore obtained each year far exceeded that for the whole country in the late eighteenth century. In 1931 out of 29,000 tons raised in Britain, 26,000 tons came from the Mill Close Mine. For many years it remained the largest lead-mine in the country. The introduction of electric pumping prolonged its life, for the workings, descending to a depth of 1,000 feet, eventually reached the water-table in the limestone. It was disastrous flooding in 1940 which unhappily brought operations to an end.

Among the few other lead-mines left, the Glebe Mine at Eyam (which is also important for fluorspar) and the new drift working near the High Tor at Matlock should be mentioned. The latter was an attempt to re-open by modern methods workings abandoned by earlier miners, but was soon closed.

There is an extensive literature on the history of Derbyshire lead-mining and its related laws and customs, but only a brief account is appropriate here in order to explain some of the interesting relics which the industry has left on the landscape. Of lead working in Roman times there is both actual and documentary evidence. Into the former category come the several pigs of lead which have been found bearing Roman inscriptions. We also know that it was exported to the homeland in Italy. Later on, the Domesday Survey mentions three lead-mines at Wirksworth, and one each at Crich, Bakewell, Ashford and Matlock, but it is strange that no reference is made to the Castleton district for it was always a rich area. The Odin Mine close to Castleton is certainly one of the oldest in The Peak.

In the Middle Ages lead production rose to a new importance and, besides its extensive use at home for piping and the roofs and windows of our cathedrals and churches, the metal ranked with wool as one of the chief exports from this country. In Derbyshire nearly the whole mining area from Wirksworth to Castleton became known as the King's Field from which a royalty was levied on the ore obtained. Mining operations were regulated by a code of laws based largely on local customs which had become blessed by the sanction of time. Many of these laws survive and meetings of the old Barmote Court presided over by the Barmaster are still held at Wirksworth to administer them.

The King's Field covered parts of both the High and Low Peak. While Wirksworth was the seat of the Barmote Court for the Wirksworth Hundred (Low Peak), its counterpart for the High Peak was at Monyash. A Barmaster held office at each place and at least during the seventeenth century several deputy Barmasters were appointed in other centres such as Matlock, Wensley, Brassington and Tissington. The Barmote Court met twice a year with a jury consisting of twenty-four experienced miners from among whom twelve were chosen for each meeting of the Court. The Court continues to meet twice yearly in the Moot Hall at Wirksworth, where the Standard Dish of the Low Peak miners is still used. The dish is a heavy oblong vessel of bronze, made in the fourth year of the reign of Henry VIII and holds fourteen pints or about 60-70 pounds of lead, varying according to the purity of the ore. There was no standard dish for the High Peak, but those commonly used at Monyash held sixteen pints. By Acts of Parliament passed in 1851 and 1852 for the High Peak and the Wirksworth Hundred respectively, the Barmote system was modified. Thus in addition to the Barmaster a steward was appointed by the Crown. The jury was reduced to twelve men, six being chosen for each court, while the standard measuring dish for both divisions

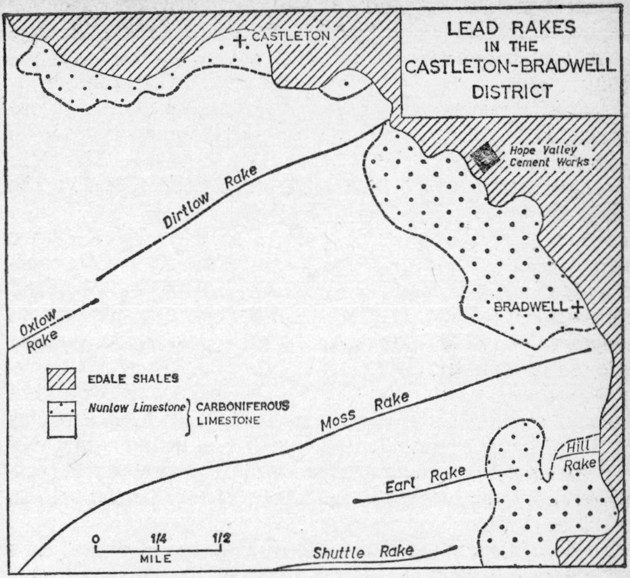

Fig. 11 Lead rakes in the Castleton-Bradwell district. (Based by permission on the Ordnance Survey. Crown Copyright reserved)

became one of fifteen pints.

Under the ancient mining laws the essential rights remaining today are those which entitle the finder of a lead vein to have sufficient land and enough water to work it irrespective of the ownership of the land. The presentation of a dish of ore to the Barmaster frees the working and thereafter every thirteenth dish is payable as a royalty to the Lord of the Field. From before the Norman Conquest until the reign of Elizabeth I the mining rights were held by the Crown and royalties were therefore paid to the sovereign, but Queen Elizabeth relinquished the rights and among those to whom many of them subsequently passed was the Devonshire family, who thus gained a source of wealth. Today the royalty, or Queen's 'lot', as it is called, is payable to the Duchy of Lancaster. Whilst it may be thought that the mining laws which were invariably upheld by the Barmote Court operated somewhat harshly against the interests of farmers and land-owners, it must be remembered that the laws were in being when much of the land was still in waste and that later owners obtained land, having knowledge of the miners' rights.

Among the miners themselves the ancient rights and customs were jealously upheld, and to the layman a good deal of mystery shrouded their activities because of the extraordinary vocabulary of phrases and technical terms which they used. Most of these terms were quite meaningless to those outside the industry. Many of them have been recorded in a poem by Edward Manlove, published in 1653. The author, who was Steward of the Barmote Court at the time, described in execrable verse the rights of the miner and the procedure of registering and working his claim. As an illustration the following extract occurs towards the end of the poem:

> Most of the Customes of the Leadmines, here
> I have described, as they are used there;
> But many words of art you still may seek,
> The Miner's Tearms are like to heathen Greek,
> Both strange and uncoth, if you some would see,
> Read these rough verses here composed by me.
> Bunnings, Polings, Stemples, Forks, and Slyder,
> Stoprice, Tokings, Sole-trees, Toach and Ryder,
> Water-holes, Wind-holes, Veyns, Coe-shafts and Woughs,
> Main Rakes, Cross Rakes, Brown-henns, Budles and Soughs.

An interesting case of the application of the mining laws occurred only a few years ago. A quarryman at Wirksworth discovered a lead vein while at work. He staked his claim in the traditional manner, in this instance to part of the quarry face. The Barmote was called and upheld the claim, and as a result the quarry-owners were obliged to pay a sum to the workman in order to regain the title to the limestone face traversed by the vein. From this it must not be imagined that all the discoveries of lead in the past were made accidentally. Very many were the outcome of patient and skilful search, often made by men who had an eye for the nature of the ground, however rudimentary their ideas of geology may have been.

RELICS OF LEAD WORKING

Mining over a long period has left characteristic scars upon much of the limestone country. In many parts the ground is strewn with hollows denoting old shafts with their attendant spoil heaps now grass-covered. Some of the longer rakes can be traced for miles by narrow ridge-like spoil banks, stretches of which are sometimes planted with trees to prevent grazing animals from being endangered by lead poisoning. Then there are the traces of early smelting. Until modern times smelting was done in primitive hearths called 'boles'. The hearth consisted of a shallow hole excavated on the brow of a hill facing the prevailing wind. The ore was broken into small pieces by men, cleaned and washed by women and girls. It was then placed in the hole mingled with wood, fired, and then covered with turf. Fanned by the wind, the heat became so intense that the melted lead flowed out into channels where it cooled and solidified into blocks called pigs which could be readily transported. The inn called the Pig of Lead in the Via Gellia may well serve as a reminder of former traffic in that area.

The bole hearths account for the frequency of the name Bole Hill in The Peak. On some of the hills bearing this name the actual hearth site is marked to this day by a patch of sterile ground. Their distribution is at first a little surprising for they mostly occur on the adjoining gritstone, especially on the East Moor and the grit-shale outliers, while only one or two are to be found on the limestone itself. The higher ground formed by the gritstone offered greater exposure to the wind and the surrounding slopes provided a plentiful supply of timber for fuel. Another advantage was that the hearths lay near to Chesterfield, which was for a long period the chief market for Derbyshire lead.

At an early date charcoal displaced wood as a fuel. It is not known when this practice began but it undoubtedly occurred while the bole hearths were still in use. Later on smelting methods were improved by the introduction of charcoal furnaces with a stone hearth and by the use of bellows to supply the draught of air. So far as is known, however, no traces of such furnaces remain.

Later still, the evidence of lead smelting is to be sought not on the hills but in the valleys by the side of the streams. This relates to a further change in technique which took place in the second half of the eighteenth century and which really marks the beginning of modern methods. From this time onwards the ore was smelted in a reverberatory furnace called a cupola, the earliest of which was erected at Ashover in 1747. The cupola consisted of a furnace containing the fire at one end, separated from the ore by a brick wall over which the flame was drawn The furnace roof sloped down towards the other end where a flue led into the chimney. The siting of these smelters on the bank of a stream enabled water wheels to operate the bellows, while the valley location favoured the use of coal and subsequently coke for firing the furnace.

With these improvements much larger quantities of ore could be treated, and the cupola served as the chief means of lead smelting during the period of maximum production in the late eighteenth century and the early nineteenth. Ruins of some of the cupola buildings can still be found. Vestiges of one are to be seen in a meadow at Alport just below the confluence of the Lathkill and Bradford streams. Two others stand in Middleton Dale just above Stony Middleton, while in other cases only a name recalls their existence such as Cupola Cottage on the roadside north of Baslow and the Cupola Footpath near Hathersage.

After the mid-nineteenth century the Matlock district became the chief area for smelting. Though further technical advances were made, the principle of the reverberatory furnace remained as the basis for the operation. One of the oldest works belonging to this period, situated at Lea, was finally demolished only a decade or so ago. The important smelting and recovery plant attached to the Mill Close Mine at Darley Dale, using the most modern methods, is still active, treating lead scrap as well as the small quantities of ore still raised in Derbyshire.

One last but important reminder of the lead-mining industry is seldom observable. This relates to the elaborate means by which many of the deeper workings were drained. In the early phases mining presented little difficulty for the ore veins outcropped at the surface. After these superficial resources were exhausted, the mines were sunk deeper and deeper until eventually the water-table was reached. Some shafts went to a depth of 200 feet before the amount of water became unmanageable The incoming water was at first removed by pumping or by small drainage tunnels. During the eighteenth century much larger underground tunnels known as 'soughs' were excavated to convey the water to the river Derwent. These were often several miles long, being cut through the limestone and lined with limestone blocks where they passed through the shales. They were works of considerable magnitude and undertaken at great cost. The first was the Cromford Sough, which was completed about 1690 at a cost of £30,000. It served to drain a number of mines in the Wirksworth district, though it brought financial loss to its promoters. Much later this sough was put to an unforeseen use when in 1771 Richard Arkwright bought it in order to supply his first cotton mill at Cromford with water power. The Stoke Sough (1732) drained mines around Eyam and reached the Derwent south of Grindleford. The Hill Carr Sough (1753) leading from the mines at Alport and Youlgreave to the river just south of Rowsley was four miles long. The Meerbrook Sough (1772) starting at Wirksworth reached the river near Whatstandwell. For the past sixty years it has supplied the Ilkeston and Heanor Water Board, yielding 12 to 15 million gallons daily.

SOME OTHER MINERALS

Around the old lead workings the accumulated waste heaps are largely made up of the other minerals that accompanied the ore in the veins. Today anyone who is so inclined can make quite a useful little collection of these minerals. The lead or galena shows as sparkling square fragments. The fluorspar occurs in clusters of cubic crystal, often honey-coloured, sometimes colourless or tinged with purple. Less abundant are the dark brown crystals of zinc blende which when broken look like resin. Barytes is common but because it is unattractive is frequently passed over. It occurs as shapeless cream-coloured masses with part of the surface covered by a rust-coloured deposit. Calcite, commonest of all, takes the form of whitish or semi-transparent crystals.

All these minerals, together with a number of others also found in the limestone, are of economic value and are worked commercially to a greater or lesser extent. Some of them, such as fluorspar, are most readily obtained from the spoil heaps and old lead workings, a few of which have been re-opened for the purpose. Fluorspar itself is now in considerable demand in modern industry. Most of the output is consumed as a flux in steel-making, while the best grade with a high fluoride content is used in non-ferrous metallurgy and in the ceramic and chemical industries. Production has given rise to renewed activity in several of the old lead-mining areas of which the chief are the Castleton-Bradwell-Eyam and the Matlock-Wirksworth districts. In recent years numbers of local men, many of them farm-workers, have been drawn to spar-winning with pick and shovel. Some of them work only at week-ends or in the evenings, while others have found it profitable to make it a full-time venture, seeking the mineral not only from spoil heaps but from old lead-mines, paying a royalty to the local landowner and selling the output to a company which owns the mineral rights. Should they actually find lead, the traditional procedure still applies, by which the claim is taken to the Barmote Court, and after the first dish of ore is made over to the Crown the mine is freed for working regardless of who owns the land.

Of special interest is the variety of fluorspar with a rich purple or amethyst colour sometimes verging to yellow. This is the well-known Blue John used for making ornaments. Its occurrence is confined to the Castleton neighbourhood, where the Blue John Mine, the chief source of supply, is situated. The deep tint of this handsome mineral is thought to be due to staining caused by the inclusion of hydrocarbons during crystallisation. Evidence for the existence of hydrocarbons is found in the small seepages of bitumen (elaterite) from the limestone, as in the Winnats Pass, suggesting the organic constituents of the limestone as a possible source. Besides the Blue John Mine, the mineral has been worked from the Treak Cliff Mine and the Old Tor Mine in the Winnats. While it is likely that this variety of fluorspar was worked much earlier, actual records of its use for ornamental purposes date only from the mid-eighteenth century. Though much skill was necessary in shaping the spar owing to its brittleness, ornaments of great beauty were made in Castleton and Buxton as well as in London. The remarkable candelabra commissioned by Robert Adam and now in the Victoria and Albert Museum, was made by Matthew Boulton of the Soho Foundry, Birmingham. Eventually exports of spar were made to France for the production of massive ormolu work and from this trade comes one of the several explanations of its popular name, for the French are said to have called it 'bleu-jaune', whence the corruption 'blue john'. It seems, however, that the mineral had received its name before this time. In recent years the finest spar with its rich amethyst colour has become relatively scarce, but the making of less expensive ornaments, obtainable in most of the Derbyshire tourist centres, continues.

Another of the vein minerals, barytes (barium sulphate) or 'heavy spar', is worked in crystalline form from old lead-mines and spoil heaps. After grinding to a fine white powder, it is used in making paint, while being heavy and chemically stable it is effective as a filling material in the paper and textile industries. Calcite (calcium carbonate) is the commonest of all the vein minerals and also occurs in limestone fissures. It is worked from the Long Rake near Youlgreave, where its uniform quality, owing to the vein being free from lead and not discoloured by other mineral matter, gives it commercial value. Ochre (red oxide of iron) is sometimes found in pockets in the limestone. Where it occurs in sufficient quantity, as in the Via Gellia and at Matlock, it has been worked as a source of colouring matter.

STONE QUARRYING

Of the major rock formations within the Peak District, the Carboniferous Limestone forms the main basis of the quarrying industry. In addition to the large-scale extraction of stone from numerous quarries, there are a few localised workings of the chert and silica deposits associated with the limestone. Quarries are unsightly. Mainly for this reason the largest centres of production around Buxton, Matlock and Wirksworth have been excluded from the National Park. The quarries remaining within the Park, are, with few exceptions, on a smaller scale. They produce less than a million tons of stone annually compared with a total output for the Peak District as a whole of well over six million tons. Tunstead (ICI), with a new cement works, constitutes the most serious future threat to the Park.

The shales and grit-shales are of no economic importance except in the Hope Valley, where the Edale Shales are worked with the adjacent limestone (Nunlow Beds) for the manufacture of cement. The resulting works form the largest industrial site in the National Park (Fig. 11). As to the gritstone, which provides relatively thin flagstones formerly used for roofing as well as massive sandstones suitable for building, its exploitation has greatly declined. Even the use of the Kinderscout Grit at Stancliffe, Darley Dale, which has a high reputation as a building-stone, has diminished, chiefly because cheaper materials are now favoured. Many famous buildings have been erected in this stone. Chatsworth House and The Crescent at Buxton are examples in Derbyshire, while St George's Hall at Liverpool provides an instance farther afield. The output of pulp-stones and grindstones, the latter chiefly for glass-grinding, continues on a small scale largely from quarries around Hathersage. Groups of abandoned grindstones, some only roughly hewn, others almost finished, can be seen by the rambler in many places as he clambers along the gritstone 'edges'. They are interesting relics of what was once a considerable industry. Such can be found along Stanage Edge, Curbar Edge and Froggatt Edge, at Burbage End and in a roadside quarry between Grindleford and Hathersage. Perhaps the best quality stone for the purpose was that provided by the Chatsworth Grit between Hathersage and Padley, hence the name Millstone Edge. Towards the end of the eighteenth century cheaper stone imported from France caused the industry to decline and eventually brought it to an end. This activity actually survived on a small scale at Tansley near Matlock, where the Chatsworth Grit again provides suitable material. Here at the Derbyshire Oaks Quarry the making of grindstone was still carried on, chiefly for use by the file manufacturers of Sheffield, although some was even exported to distant countries such as India, but the quarry was closed in 1970. A new use for some of the discarded stones has recently been found as boundary signs on the main roads at points of entry to the National Park. Mounted edge-wise on a massive stone pedestal, they afford an appropriate emblem.

Considering the extent of the gritstone surface, which forms the entire northern moorland as well as the flanking East and West Moors, the number of active quarries is remarkably small compared with even fifty years ago. In fact abandoned quarries, moss-covered and partly overgrown with vegetation, are a typical feature of the gritstone country. They are almost as numerous as the old houses and farm buildings which were built from the stone they supplied. The decline in the demand for stone, except in small quantities, is in marked contrast to the ever expanding demand for limestone and it is unlikely that gritstone quarrying will ever again be required to meet the needs of great enterprises such as railway building and reservoir construction which brought it prosperity a few generations ago. High quality gritstone is still worked on a small scale at Stanton Moor and elsewhere.

In the east on the outskirts of Sheffield, yet within the National Park, the clays of the Lower Coal Measures are worked for fireclay. These clays are siliceous and withstand high temperatures, so that they are used for the manufacture of firebrick and furnace linings. The workings, which are near Bradfield, are small and, being chiefly in the form of adits, do not seriously disfigure the surface. A similar deposit of fireclay is worked in the extreme west on Bakestonedale Moor. From beneath the clay in both localities, gannister, another refractory material, is obtained. Old coal shafts which are still recognisable on Bakestonedale and Axe Edge Moors were in production about a hundred years ago but the coal was poor and very sulphurous. As Miss Nellie Kirkham reminds us, perhaps the most interesting feature of these shafts was their names: Black Clough, Wooden Spoon, Crash-away.

LIMESTONE IN ABUNDANCE

To return to the limestone. To the modern chemical industry in its various forms the Carboniferous or Mountain Limestone of north Derbyshire is of great importance. In the first place, its purity in terms of the calcium carbonate content is exceptionally high and stone of this degree of purity exists here in greater quantity than elsewhere in the country. Moreover, its mode of occurrence favours large-scale methods of quarrying, crushing and burning, while its location is within a short distance of the principal consuming centres in south Lancashire and Cheshire and the West Riding of Yorkshire. These advantages are greatest in the Buxton and Miller's Dale district, where the Chee Tor and Miller's Dale Beds yield high quality stone to a depth of several hundred feet. This is the area of huge quarries including the Tunstead Quarry with its enormous working face nearly a mile long, which is by far the largest in the country. The distribution of the quarries was originally determined by accessibility to the railways, and as time went on a confusion of branch lines arose threading their way into the works from the main lines. Over 90 per cent of the lime produced in the Peak District now comes from this area. The same beds are also worked at Stony Middleton, while the other centres of limestone production include Bradwell and Eldon Hill, together with Matlock and Wirksworth in the south.

Of the total output of lime from the works in The Peak, nearly 40 per cent is consumed by the chemical and allied industries. Some 20 per cent is used in the preparation of agricultural fertilisers and about 12 per cent in the building trade. Large quantities of stone are used as a flux in iron and steel production and also as road metal. Much of the output from within the National Park is quarried for road metal which reflects the poorer quality of the stone available except of course that found at Stony Middleton. During the nineteenth century and earlier, however, most of the stone was burned to provide lime for the fields and it is impossible to go far in this part of Derbyshire without seeing vestiges of the numerous kilns erected for that purpose.

Limestone-working, whether upon the plateau or in the dales, affects the amenity of the countryside. Apart from the disfiguring effect of the quarry face, smoke from the kilns makes the working all the more conspicuous and may also become a nuisance, while the coating of fine grey dust which covers the surrounding vegetation, sometimes stifling its growth, is particularly distressing. Unfortunately the situation is likely to become worse rather than better, for the demand for lime continues to increase, besides which some of the large quarrying concerns have reserves of limestone lying inside the boundary of the National Park. This is the sort of issue which may one day have to be faced squarely if national parks are to fulfil their purpose, although nobody should suppose that amenities can be preserved entirely without cost. Recognising the importance of this matter, the Peak Park Planning Board issued a statement of policy in 1952. This made clear that future quarrying projects would receive the closest scrutiny and that where possible operation should take place outside the Park. In any event the Board reserved the right to exclude quarrying altogether from certain parts which are at present quite unspoiled and are of the highest scenic value.

Before leaving the subject of limestone, reference should be made to two special varieties which have been used for purposes quite different from those already mentioned. One is the Hopton Wood Marble and the other the once famous Ashford Black Marble. Neither of them is a marble in the strict geological sense, though both are decorative stones and take a high polish. That from Hopton Wood, south of the Via Gellia, is a cream-coloured fossiliferous stone belonging to the Chee Tor and Miller's Dale series, which affords a valuable building material chiefly for interior work. The chief source of this fossiliferous stone is now the Steeplehouse Quarry at Wirksworth. A recent development at Hopton is the extraction of magnesium from dolomitized limestone, an industry which has rendered the country largely independent of imported supplies of this light metal alloy. To those interested in historic building-stones, the old quarries in Ricklow Dale near Monyash, whence came the famous Derbyshire grey 'marble' in the eighteenth century, are worth seeing. Even their eerie desolation is impressive.

AN ELEGANT MARBLE

At Ashford-in-the-Water near Bakewell traces can still be found of the Black Marble workings of over a century ago. This marble was obtained from beds of the limestone containing carbonaceous and clayey matter. It provides a handsome material for ornamental, including inlay work, and was once the basis of an important local industry now largely forgotten. It was certainly worked in the sixteenth century, and in 1748 a marble mill was set up in Ashford village, where the industry survived until about 1905. Far earlier, however, is the first evidence of the use of Black Marble, for a dressed slab of it was found in a prehistoric burial site in the tumulus on Fin Cop, the hill about two miles north-west of Ashford. In the immediate neighbourhood of Ashford old underground workings can be traced in the Rookery Plantation just west of the village and opposite on Arrock Hill to the south of the Wye.

OTHER QUARRY PRODUCTS

Associated with the Carboniferous Limestone are several other materials of commercial value which contribute to the quarrying industry. Chert, a siliceous mineral closely resembling flint, which occurs locally in the limestone, is worked near Bakewell for use in the Staffordshire potteries. One of the two remaining sources, the Holme Bank quarry, was first opened by Sir Josiah Wedgwood nearly two hundred years ago. Silica deposits, forming large pockets in the limestone, form the basis of an importance industry producing refractory materials. These pocket silicas, as they are called, are found near Friden and Parsley Hay to the east of the Ashbourne-Buxton road. They are composed of brightly-coloured sands, gravels and clays. The largest of them are several acres in extent and reach to a depth of over 100 feet. The workings are therefore in the form of pits. To geologists the origin of these deposits has proved puzzling, although more is now known about them than was the case even a decade or so ago. It is agreed that they are accumulations filling large sink-holes in the limestone. In the view of Dr P. E. Kent they are mainly of Mesozoic age and represent the subsided remains of former widespread sediments of Permo-

Triassic times. There is good reason to suppose that the deposits have been preserved by sinking progressively farther down as the sink-holes deepened through later geological times. Some limestone-shale (Pendelside) material also occurs, relict from a period before the over-burden was stripped from the limestone, while the latest deposits are represented by clays containing silicified wood of Tertiary age and by patches of Boulder Clay, each member of the succession being preserved beyond the normal limits through continuing local subsidence.

The various igneous rocks which are found in the Carboniferous Limestone of The Peak are chiefly used as road-making material. Some of the dolerites and basalts provide very durable stone of high crushing strength and are therefore suitable for roads bearing heavy traffic. The chief centre of production was at Calton Hill near Taddington, where the quarrying of basalt, begun in 1906, was considerably extended in recent years, but ended in 1969. Here the amount of waste caused by the presence of volcanic ash with the basalt, is particularly great, so that ugly spoil heaps now surrounding the hill occupy a large part of the workings. Another source of basalt is from the Waterswallows Quarry north of Buxton. Clearly the good roads of the Peak District owe much to the availability of local materials. Were it not for these, suitable stone would have to be brought from more distant areas such as Charnwood Forest or the Lake District.

REFERENCES

Hopkinson, G. *The Laws and Customs of the Miners within the Wapentake of Wirksworth.* (1644)

Tapping, T. *Treatise on the High Peak Mineral Customs and Mineral Courts Acts, 1851 and 1852.* (1854)

Stokes, A. H. Lead and Lead-mining in Derbyshire. *Trans. Chesterfield and Derbys. Inst. of Engineers: 8 (1880-81) and 9 (1881-92)* (1854)

Thorpe, H. Derbyshire Lead Production. *Derbyshire Countryside* (Jan, 1934)

Traill, J. G. The Geology and Development of the Mill Close Mine, Derbyshire. *Econ. Geol.: 34* (1939)

Luggard, C. E. Derbyshire Lead-Mining Customs in the Sixteenth Century. *Jour. Derbys. Arch. and Nat. Hist. Soc.: LXV* (1940)

O'Neal, R. A. H. *Derbyshire Lead and Lead Mining: A Bibliography,* 2nd ed. Derbyshire County Library, Matlock (1960)

Raistrick, A. and Jennings, B. *A History of Lead Mining in the Pennines.* Longmans (1965)

Fuller, G. J. Leadmining in Derbyshire in the Mid-Nineteenth Century. *East Midland Geographer: 23* (1965)

Harris, H. *Industrial Archaeology of the Peak District.* David and Charles (1971)

13. Water for Cities

Pure water is the best of gifts that man to man can bring.
Attributed to the Hon. G. W. E. Russell

The traveller going by road from Sheffield to Manchester across the moors of the High Peak finds, by whichever route he takes, that it skirts the shores of one or more lakes. If he chooses the road via Woodhead he will meet two lakes in the Don Valley beyond Stocksbridge and later, on descending the valley of the Etherow (Longdendale), his route will follow a chain of lakes for at least five miles. If he takes the road leading over the Snake Pass, he will find on dropping to the Derwent Valley that it follows the edge of a body of water larger than any of the others. It is not an exaggeration to say that The Peak is a lakeland, but it is an artificial one created by the construction of reservoirs in many of the valleys. Except for a few examples, these reservoirs are too numerous to mention by name, but altogether they impound a vast quantity of water for public supply. Water is in fact one of the principal resources of the Peak District, and the supplies are organised mainly to meet the needs of the urban population in the bordering lowlands to the east and west and even farther afield. Quite apart from industrial consumption, about four million people are wholly or partially dependent upon these sources.

Most of the water authorities store their supplies in surface reservoirs which provide the northern part of The Peak with its lakes. These are placed so that they catch the natural drainage from the surrounding slopes over as large an area as possible, this area being termed the catchment or gathering ground (Fig. 12). The majority of the reservoirs and their gathering grounds lie within the National Park, and altogether they cover some 160 square miles of territory or 29 per cent of its total extent. Over this considerable area the water authorities have been the means of altering the scenery. Not only have they introduced the lakes, but in taking measures to prevent contamination of the water they have established many coniferous plantations and have been able to impose restrictions on agricultural activity. These changes alone have created what might be regarded as a distinctive water-supply landscape. Other changes, too, have taken place. As a result of building some of the larger reservoirs, farms and even villages have disappeared, while roads have had to be diverted or new ones made.

SURFACE STORAGE ON THE GRITSTONE

The reservoirs and their gathering grounds are confined to the Millstone Grit country, for the grits and shales which form the surface are generally impermeable and ensure an abundant runoff. As a contrast the absence of reservoirs on the Carboniferous Limestone reflects the permeability of that rock and the relative scarcity of surface water. The problems of water supply in the limestone country are of special interest, however, and will be mentioned later.

Physical conditions in the gritstone area are especially favourable for the large-scale organisation of water supplies. They include a high mean annual precipitation of 40 to 60 in. evenly distributed through the year; a low rate of evaporation favouring the use of open reservoirs; and large tracts of impermeable rock dissected by narrow valleys suitable for the building of dams. While the gritstone is impermeable in a general sense, the upper layers forming the surface are often somewhat porous as the result of weathering. This produces an absorbent layer, often thickened by overlying peat, which serves to regulate the drainage and to balance seasonal inequalities in the rainfall. Moreover, the water from the gritstone is soft, and while this is a general advantage it is particularly so in the case of many industrial consumers.

The demand for water from the Peak District developed with the growth of industry and population in the nineteenth century. By the 1830's, when the first reservoirs were built, many of the valleys already contained small mill-dams serving textile mills. The early water undertakings were therefore obliged to provide what is called compensation water to meet the needs of the mill-owners. This meant that their reservoirs had to be designed to hold a larger quantity than they themselves required. The provision of compensation water has become a statutory obligation and it can often be a significant item in the planning of a new supply scheme, as some of the later ones in The Peak have shown.

Among the first of the large authorities to draw supplies from the moors was the city of Sheffield, the Redmires (1836-54)

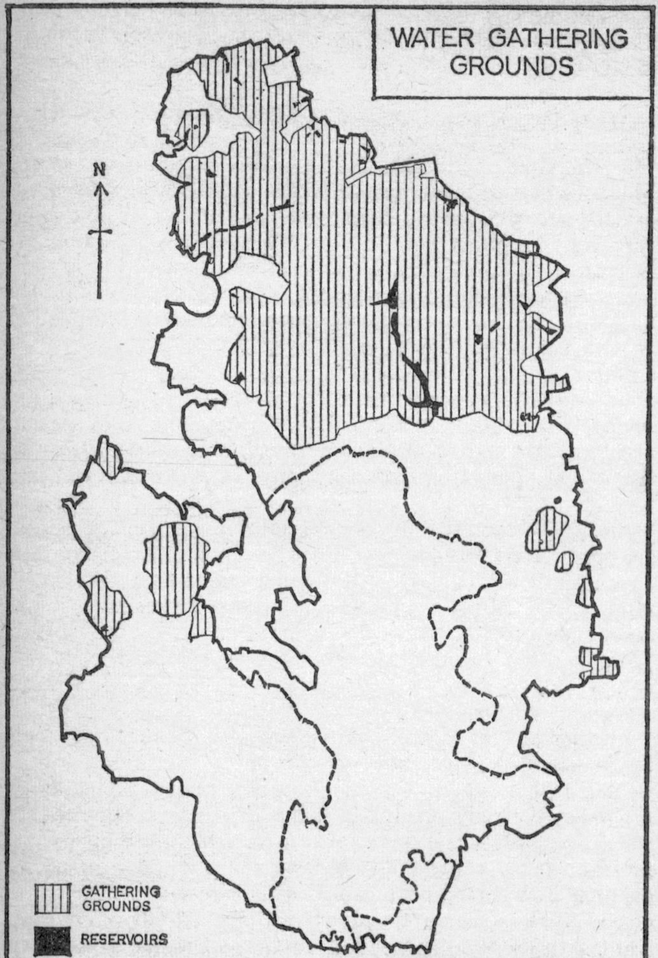

Fig. 12 Water-gathering grounds in the National Park

and the Rivelin (1848) group of reservoirs all lying within six miles of the town centre. Manchester's chain of reservoirs in Longdendale were larger. They were begun in 1848 and completed in 1862, and now have a total capacity over 3,000 million gallons.

While the principle of each large town utilising its nearest upland valley became generally recognised, some poaching by the more powerful authorities was inevitable. In any case the larger undertakings were eventually forced farther into the interior of The Peak. Thus Sheffield, Barnsley and Huddersfield all turned to more distant sources, while Manchester, frustrated by similar developments from Stockport and Ashton-under-Lyne, turned to the Lake District and after 1877 built no further reservoirs in The Peak. Stockport, however, continued to do so by using the Kinder stream for the Kinder Reservoir (1912) and the Goyt for the Fernilee Reservoir (1937).

In 1899 the Act authorising the formation of the Derwent Valley Water Board, representing a joint project to supply Sheffield, Derby, Nottingham and Leicester, made it possible to draw upon the hitherto untapped sources of the upper Derwent and its tributaries. This scheme, one of the largest in the country, has been developed to its present extent in three stages. Firstly the Howden and Derwent Reservoirs, completed in 1912 and 1916 respectively, impounded supplies from the Derwent headstreams; secondly, water was diverted from the Ashop and Alport streams to the Derwent Reservoir, increasing its capacity by nearly 50 per cent; thirdly, the great Ladybower Reservoir, begun in 1935 and completed in 1943, impounds water from the lower Ashop and several small left-bank streams, including the Ladybower Brook, flowing from the Derwent Moors. This reservoir alone holds 6,300 million gallons and is the largest in Britain. Its construction involved the submergence of the small villages of Derwent and Ashopton. Before this part of the valley was inundated, Derwent Hall, a seventeenth-century house, was demolished and the old pack-horse bridge across the river was removed. Happily it was rebuilt at Slippery Stones, higher up the river, as a memorial to John Derry, at one time editor of a Sheffield newspaper, whose handbook, *Across the Derbyshire Moors* (1904), did much to acquaint the general public with the byways of The Peak. A further stage, undertaken more recently, was the bringing of supplies from Edale by diverting water from the River Noe and the Jaggers Clough stream by means of an aqueduct through the hills to the Ladybower

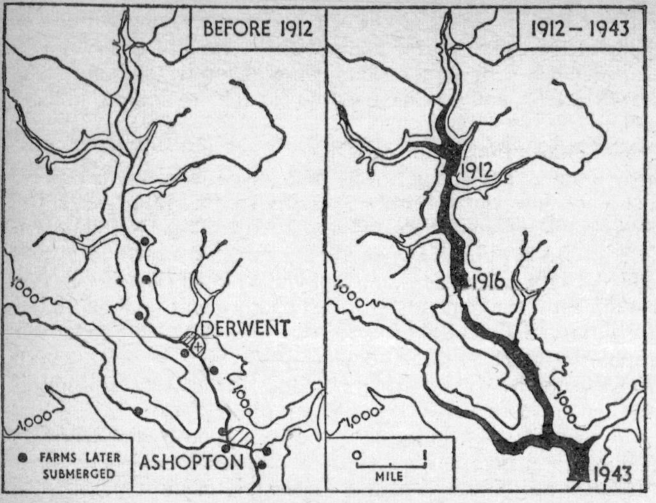

Fig. 13 The Upper Derwent Valley before and after reservoir construction. (Based by permission on the Ordnance Survey. Crown Copyright reserved)

Reservoir (Fig. 13). With regard to the last-named reservoir, an incident of unusual interest occurred during the exceptionally dry summer of 1959. During this season the water level fell so low that a farmer was able to recover stones from the farm which he had worked before the reservoir was constructed some sixteen years previously. His intention was to use the stones for a rockery at his new home in Ashbourne as a reminder of the property alongside the Derwent which his family had occupied for at least a century.

The function of the Derwent Valley Water Board is to provide bulk supplies to the towns already named and to various authorities in the county of Derbyshire. After meeting the needs of the county undertakings, which are relatively small, the remainder is made available to the four towns in the following proportions: Sheffield 25 per cent, Derby 25 per cent, Leicester 35·72 per cent, and Nottingham 14·28 per cent. The small share apportioned to Nottingham is due to the fact that the city already possesses other large-scale sources. After making adjustment for sales and purchases of water under inter-authority agreements permitted by the Act of 1899, the actual proportions are now as follows: Sheffield 29·7 per cent, Derby 19·3 per cent, Leicester 36·5 per cent, and Nottingham 14·5 per cent. The Board now supplies about 44 million gallons daily of which the four towns together take 41 million gallons. From the reservoirs the water is conveyed by aqueducts. The supply for Sheffield passes directly through the Rivelin Tunnel, while that for the other towns, after being filtered at Bamford, is conveyed southwards by an underground aqueduct over 50 miles long which crosses the Trent at Sawley on its way to Leicester. From Ambergate a branch leads to Nottingham and from Little Eaton another goes to Derby. The course of the aqueduct can be traced in many places by the inspection chambers seen at intervals on the surface. At first sight these structures with stone walls and red-tiled roofs often puzzle the hiker as he makes his way along the east side of the Derwent Valley. The most recent completion of a reservoir was that at Errwood in 1967 by the Stockport Water Authority, and, in accordance with some modern practices, sailing is permitted on it.

From what has been said it should be evident that the northern moors are the most intensively used for water conservation. In this area almost the entire surface is devoted to catchment, which of course reflects the heavy demand on both flanks of the Pennines. Comparatively little water is impounded in the West Moors because of the smaller requirements in the adjoining lowland. Rudyard Lake near Leek, authorised in 1797 as part of the Trent-Mersey Canal system, was originally constructed to supply this Canal and some of its branches. On the East Moors, too, despite the demands of the Chesterfield industrial district, large reservoirs are lacking. This is partly due, however, to the absence of suitable sites, for the valleys of the small consequent streams draining eastwards have very limited catchment areas, making the cost of reservoirs high in relation to their capacity.

While reservoir construction has proved comparatively easy in many of the valleys, such work has not been without its problems. Instability of the rock, in particular, resulting either from faulting or from slipping, such as sometimes occurs at the contact between grit and shale, has increased the cost of some of the dams. In the earlier phases of the Derwent scheme, for example, the shattering of the weaker shales by pressure from beds of overlying grits necessitated a further excavation of the valley bottom to secure a watertight floor for the Howden and Derwent Reservoirs. Again, the Broomhead Reservoir in the Ewden valley, belonging to Sheffield, though completed in 1929, could not be brought into use for another five years owing to landslips on one of the banks and leakage through underground fissures. The absence of reservoirs in some apparently favourable valleys, despite the increasing demand for water, can often be explained on grounds of geological instability. The zone of landslips on the north slope of the Woodlands Valley has so far deterred the Derwent Valley Water Board from using that site.

THE WATER SUPPLY AUTHORITIES

While most people would agree that the reservoirs have improved the scenery of The Peak, the reservation of large tracts of land for water conservation poses an intricate problem for the National Park. This is because the protective measures adopted by the supply authorities over a wide area have led to restrictions of varying degree upon public access as well as upon agricultural activity. The issue is really one of competing claims made upon the same land by three different interests: water supply, agriculture and public recreation. Since the demands of water supply are already great and are likely to increase, it seems desirable to consider in a little more detail the technical basis of the protection policy and the nature of the restrictions to which it gives rise.

The responsibility of the water authorities is to provide a reliable and wholesome supply. The purity of the water is of vital importance and this must be scrupulously safeguarded. This is done in three different ways: firstly, by controlling all activities within the gathering ground in order to reduce the risks of contamination; secondly, by storing the water in the reservoir, during which purification is further assisted; and thirdly, by filtration and chemical treatment prior to distribution. The main risk of contamination, however, is through infected members of the human or livestock populations, so that great importance is attached to the protection of the gathering ground. If, for example, typhoid bacteria were conveyed to a reservoir, storage for a period of seven days would normally suffice to render them harmless, but such a period cannot be relied upon in all reservoirs at all times of the year. Even though chlorination has been compulsory since 1939, the possibility of a failure in the various safeguards has caused the authorities to maintain a rigorous policy of protection at the source. The remote chance of such a failure was the main submission of the Huddersfield Corporation some time ago in objecting to a proposed section of the Pennine Way footpath near the Digley Reservoir. In this case the Ministry concerned did not support the water authority.

Although open moorland is preferable to agricultural land for water catchment, the demand for water in the Peak District has been so great that many authorities have been obliged to include cultivated land in their gathering grounds. Some have actually acquired the entire area draining to a reservoir and after removing all farm buildings have planted coniferous trees. Others have purchased the land immediately around the reservoirs and feeder streams, thus securing control over the most important part of the gathering grounds. In all, the leading undertakings now own nearly 50,000 acres within the National Park. At least a half of this amount is open moorland over which public access is restricted. Control has been further strengthened in recent years by legislation, both through individual Acts promoted by the water authorities and by the Public Health Act of 1945. Thus cattle grazing is now prohibited near the reservoirs, while restrictions on grazing in any part of a gathering ground are now often included in farm tenancy agreements. Sheep, largely because they do not break up the sward to the same degree, are permitted to graze as far as the reservoir fence. Cultivation adjacent to a reservoir is also prohibited, chiefly to prevent solid matter containing mineral salts from reaching the storage water.

As far as public access is concerned the conflict of interests is again difficult to resolve. While the danger of contamination resulting from large numbers of people traversing the gathering grounds is undoubtedly real, the denial of access to the open moorlands of the National Park presents something of a contradiction in terms. Often, it is true, the restrictions imposed by the water authorities are indirect rather than direct. Thus the setting out of plantations, which assist in regulating the flow of water and at the same time afford a productive use of land, sometimes has protection as its primary aim. There have also been cases of the water authorities securing the withdrawal of catering licences from refreshment places in the gathering grounds. While such measures inevitably cause irritation to some sections of the public, the need for most of them can be justified. But can all of them be justified? Whatever the answer, it is surely reasonable to suggest that if freedom of movement must be curtailed to ensure the provision of a vital service, the National Park authority should do its utmost to make full and generous provision in alternative areas for the enjoyment of open-air recreation.

UNDERGROUND SOURCES IN THE LIMESTONE

Water supply in the limestone area is quite a different matter. As we have seen, conditions there do not lend themselves to large-scale surface conservation. The problem is therefore really one of satisfying the needs of the population in the area itself. Local supplies are not more difficult to find than on the gritstone but they are more liable to fail in dry seasons. Springs and wells are the traditional sources on the limestone and have been the means by which villages and farms have existed from the earliest times.

It has already been shown that local water supply was a fundamental factor in accounting for the distribution of settlements. Since so much of the drainage is underground many of the deeper wells have proved unfailing. That their importance has always been recognised by the communities which have benefited from them is shown by the interesting custom of well-dressing. This is still observed in a number of villages and consists of an annual thanksgiving service held at the well, which is elaborately decorated for the occasion. This custom is held by some to be of pagan origin, though how far back it can be traced in Derbyshire is not really known. It has been observed in many places at different periods, notably at Tideswell, Tissington, Youlgreave and even Buxton. In some it has been revived after a long lapse. At Tissington the ceremony dates back for several centuries, perhaps to the great drought of 1615. In that year no rain fell between March 25th and August 4th, except for one shower during May, so that Tissington must have appeared as an oasis in a thirsty land, for its wells did not fail. On the well-dressing occasions each well is covered with floral decorations to which are added panels sometimes in triptych form showing scriptural scenes, painstakingly executed in leaves, moss and innumerable petals of garden flowers mounted on a surface of damp clay.

In the past few years some villages, including Brampton and Barlow, near Chesterfield, which are not on the limestone, have revived the custom, though in these cases the decoration of the wells as an attraction to visitors conveys at least a suggestion of commercialism.

Many of the limestone springs have been famous in the past, and indeed still are, owing to some curious feature of their behaviour or quality. Some are noted for their intermittent character. Of these the best known was the Ebbing and Flowing Well above Barmoor Clough, near Chapel-en-le-Frith. Its alternating rise and fall occur more frequently in winter than in summer and in dry weather only a patch of damp ground marks its position. It is popularly thought that Tideswell derived its name from a spring of this nature but this is not the case, for the first syllable refers neither to tide nor time but to a person. The name means Tid's well or stream. Other springs are famed for their petrifying properties like those at Matlock Bath, while some, quite apart from the thermal springs, have been known for their supposed healing qualities. All these and many others are interesting phenomena relating to underground drainage in the limestone country, but they play little part in the provision of public water supply.

The presence of beds of lava in the limestone is a valuable asset to villages and farms in the higher parts of the uplands. These layers, being impervious to water, prevent much of it from running underground and hold it up to form elevated natural reservoirs which can be tapped by shallow wells or from springs which issue forth where the layer outcrops high up on the side of a valley.

Other means have been adopted for tapping underground supplies in the area besides the use of springs and wells. There is evidence that about the year 1700 the eminent engineer, George Sorocold of Derby, provided Wirksworth with water pumped from an old lead-mine near the town which had become flooded. Closely akin to this method is the use now made by the Ilkeston and Heanor Water Board of the Merebrook Sough, which drains a group of mines in the same district. In this case, however, the supply is conveyed outside to serve a distant area. The commonest method of supplying the larger settlements of the limestone country today, however, is by pumping water from the principal streams and distributing it from service reservoirs. In this way a piped supply has gradually been developed which now serves almost all the villages. In the Bakewell Rural District this system is now being extended to include small hamlets and some outlying farms. Similarly a quantity of water is pumped from the limestone at Bradwell to the Ladybower

Reservoir.

The provision of an adequate and regular supply of water to the scattered rural population of the Peak District is naturally expensive, but improvements are steadily being made in most parts. Beyond the limestone area the Hope Valley and the Upper Derwent are already well served, but further north, even in Edale, piped supplies are lacking because of the sparse and scattered population. Here indeed is a paradox, for this is the region of large-scale conservation – water in abundance but organised for export. Improvements elsewhere, however, should bring important benefits to rural life for water is a necessity of civilised living. Along with other essential services such as electricity, water supply contributes to the stabilising of comparatively remote agricultural communities by lessening the motives for emigration. Moreover, as far as the National Park is concerned, the extension of supplies is equally important, for a wider distribution will make the area all the more attractive to visitors from outside.

REFERENCES

Stephens, J. V. *Wells and Springs of Derbyshire*. H.M.S.O. (1929)

A Short History and Description of the Undertaking. The Derwent Valley Water Board (1945)

Porteous, C. *The Beauty and Mystery of Well-dressing*. Pilgrim (1949)

Ministry of Health Reports on Regional Water Supply Surveys, in particular:
Hawkesworth, I. H. Derbyshire and South Nottinghamshire. (1950)
Hawkesworth, I. H. The West Riding of Yorkshire. (1950)

Lockyer, A. G. A Study of Water Supply in Derbyshire. *East Midland Geographer: 8* (1957)

14. Mills and Factories

> Since what by Nature was denied
> By art and industry's supplied.
> *Samuel Butler: Upon Plagiaries*

It was explained in the first chapter that the larger industrial centres of the Peak District were deliberately excluded from the National Park. As a result, apart from quarrying, industrial activity within the Park is restricted to scattered small-scale undertakings. The exception is the cement works in the Hope Valley, which employ about 500 people. This is by far the largest manufacturing unit and is the only one which seriously obtrudes and gives a jarring note to the landscape, though the quarries on Eldon Hill and in Stony Middleton offend only a little less. From the economic standpoint, however, the location of this industry is highly favourable. The basic materials, limestone and shale, are immediately at hand and the large quantity of water required is also available. A short branch line from the main railway provides for the movement of coal, which is brought from collieries in east Derbyshire, not many miles away. This is an important item, for some six or seven tons of coal are consumed in the production of one ton of cement. Prior to the opening of these works in 1929, there was only one other cement-making plant in the north of England. Although cement is a bulky product, when packed in bags it is most conveniently distributed by road transport at least for distances up to forty miles from the factory. Within such a radius from the Hope Valley works lie the industrial districts of South Lancashire and the West Riding, so that these areas provide a valuable market. In addition to the amount carried by a fleet of nearly a hundred vehicles, considerable quantities are conveyed by rail.

Now there is no doubt that industrial undertakings are gradually becoming more responsive to the need for avoiding unnecessary disfigurement of the countryside, and the case of the Hope Valley works is of special interest in this respect. When the project was first made known, a vehement outcry not unnaturally greeted the threat to the valley, and when cement production actually began some of the worst fears were realised, as was inevitable with such an industry. This is not the place to pass judgment as to whether the choice of site was right or wrong; in any case it was made long before the present planning legislation came into being. Moreover, whatever the effects upon the surroundings, it can be amply shown that the industry afforded timely relief to local unemployment and instilled new life into the district. Eventually the company owning the works yielded to the pressure of public opinion concerning the effect of its quarrying and processing operations. In 1942 it agreed to draw up a plan for development which should safeguard the amenities of the area as far as possible. For this purpose a distinguished architect and landscape designer was appointed. The report accompanying his plan recognised the conflict of interests in the following words: 'On the one hand is an historic landscape and on the other a highly productive industry; and both are of national value.' Since that time the company has acted on various suggestions made in the plan. Even earlier, however, some measures were taken to deal with the enormous quantities of fine dust created by cement-making. At considerable cost equipment was installed by means of which over 90 per cent of the dust was trapped before it escaped into the atmosphere. The remaining 10 per cent, however, continued to be a source of annoyance, so that a few years ago the height of the works chimney was raised to 400 feet, enabling the rest of the dust to drift away at an elevation well above the village of Hope and its neighbourhood. A more recent expansion of the works has resulted in the dismantling of the chimney and the erection of a new one over 600 feet high. The firm, on the other hand, gave 40 acres of land at Overdale, Bradwell, to the Derbyshire Naturalists' Trust in 1971. Thus the filling of the old Hadfield Quarry with waste from a new one and the restoration of disused clay pits by tree-planting have helped to reduce the disfigurement of the surface.

There is an even larger industry which although not included in the National Park, is so close to the boundary as to make little difference. This is the Ferodo Works producing brake-linings at Chapel-en-le-Frith. The story of its development is extremely interesting for it typifies the achievements of the numerous individuals possessing energy and initiative to whom the Peak District has given birth. The founder of the industry was Herbert Frood, who lived at Combs, a village near Chapel just within the Park boundary. In his younger days Frood was a footwear salesman and on his travels through The Peak he often gave thought to the problem of the braking of carts on steep hills. The idea of a brake-lining occurred to him and as a result of experiments carried out in a small shed in his garden, aided by power from a water-wheel, it was not long before he invented a brake-shoe. By 1897 Frood's brake-linings for horse-drawn vehicles were in large demand and he secured part of an old mill at Chapel for making them. In a few years the dawn of the motor age gave him new opportunities. With the successive improvements which he made to brake and clutch linings, output expanded rapidly, and in 1920 a new factory was started on the site of the present works. The latter employed nearly 3,000 people and are said to be the largest in the world producing friction linings. Here is an example of an industry which attained national importance in a few decades. Its initial success was due to the personal qualities of its founder and to the skill of the workers he recruited from the immediate neighbourhood. Yet almost all the raw material required, such as rubber, asbestos, gums and resins, must be imported from abroad. On the other hand, besides serving the home market, the firm has developed a large export trade. Ferodo – an anagram on Frood's name – originated from a challenge afforded by the Derbyshire hills and the same hills today serve as an admirable testing ground for its products. The firm now employs fewer workers at Chapel owing to the opening of two factories elsewhere.

TEXTILE MILLS PAST AND PRESENT

Most of the small mills and factories within the National Park are discreetly located in a few villages or are hidden in the deep dales. The chief form of industrial activity is textile-working. Ever since Arkwright demonstrated the value of water as a source of power at Cromford in the eighteenth century, first by using the discharge from the Cromford sough in 1771 and a few years later the Derwent itself, cotton-spinning has been a characteristic Derbyshire industry. Following the success of Arkwright and his partner, Strutt, many mills sprang up on the banks of Derbyshire streams, and for a time this industry was as important here as in Lancashire itself. J. Farey, writing in 1811, noted more than 120 cotton mills in the county. The old mill in Edale is interesting in this connection. It was built in the seventeenth century as a corn mill and was later used as a tannery; early in the nineteenth century it was converted into a cotton mill and did not finally cease production until 1934.

Although it was formerly more widespread, textile processing survived in the valleys of the Derwent and Wye. Cotton doubling was carried on at Bamford and at the Cressbrook Mill in the Wye Valley but both are now closed. At Litton Mill, farther up the Wye, nylon and terylene are worked. Two small factories in Bakewell produce knitted wear, while near the south-western boundary of The Peak two dyeworks operate on behalf of silk manufacturers in Leek.

Most of these mills originally required water both for power and for processing, but today their machinery is driven by electricity. The Wye and the Derwent, however, are both used for generating electricity on a small scale, the Wye more than the Derwent, because the great storage reservoirs restrict the flow of the latter. Bamford Mill produced a little power from the Derwent and the Wye was used for the same purpose at Litton and Cressbrook Mills as well as at Bakewell, where a factory making electric storage batteries long obtained power from this source. It originated in 1777 as one of Arkwright's cotton mills, but the battery industry dates only from 1898. Until quite recently its two great water-wheels were a familiar sight to people passing by on the Buxton road. They were removed in 1956 when new generating equipment, still using the river, was installed. Except for a short period in the summer, the current obtained in this way provides a useful supplement to the main supply. This contribution depends upon an earlier impounding of the Wye upstream, which forms the 'Ashford Lake', together with the large old mill pond (Fig. 10). The battery works closed in 1970 and gave place to a Sheffield cutlery firm.

Some of the mills, the one at Cressbrook for instance, built of stone on simple and well-proportioned lines, are dignified and even elegant in appearance, and they remain as satisfying examples of the industrial architecture of an earlier age. The cotton mills in particular, however, face problems of labour supply, high costs of power, and distance from the main spinning centres in Lancashire, so that it is not easy to predict how much longer they will persist.

VILLAGE FOOTWEAR FACTORIES

At Eyam and Stony Middleton, close to the Derwent Valley,

footwear production, established over a century ago, remains active. Light shoes are made at Eyam and heavy boots for miners and quarrymen at Middleton. Of the four little factories in the latter village, the two smallest still produce hand-made boots, while the two larger ones each have an annual output of some 25,000 pairs of machine-made heavy boots, most of which are sold in the Derbyshire coalfield. Here again the problems of labour supply and production costs are apparent, the employment position being made more difficult in recent years by competition from the local limestone quarries and the fluorspar mine at Eyam. The making of shoes at Eyam, on the other hand, is favoured by the considerable use of female workers of whom there is no shortage. In fact it was this circumstance which led a Leicester footwear firm to start a branch factory here only a few years ago, although production has since declined.

NEW INDUSTRIES

The other industries located within the Park are generally of recent origin. Most of them were introduced as part of the dispersal policy during and just after the last war, chiefly as branch units of firms in Sheffield and Chesterfield. These have certain features in common: they are flexible as regards location, they use largely unskilled or semi-skilled labour, including a considerable proportion of females, they all use road transport and have no power requirements other than electricity. None, moreover, requires so specialised a type of building as to make the adaptation of old ones impracticable. Disused premises in or near the Derwent Valley offered the best prospects to such firms, especially as the area lay within daily travelling distance from Sheffield. As a result light engineering was introduced at Bradwell, a steel-hardening and polishing works at Brough, the making of metal fittings in the old cotton mill at Calver and also at Bamford. Other forms of engineering were brought to Bakewell and Tideswell, the latter place also being chosen for the making of plastic articles and for a separate plastics research unit. At Brough two industries related to the Hope Valley cement works have appeared. One is the manufacture of pre-cast concrete products, using cement from the large works and limestone from the Eldon Hill quarry, and the other the making of paper bags used by the cement works and the local light engineering firms.

It is clear that the recent phase of small-scale industrial development owes much to the availability of buildings. In this respect the pattern of distribution closely resembles that of the old textile mills, which in turn were dependent on water-power sites. Some large villages, like Youlgreave and Winster, which are not situated on major streams and which had no disused premises to offer, have not attracted new industry. It is also significant that a primary reason for the persistence of the new industries since the war is the amenity value of their surroundings. Given good communications, the move from the crowded city has been appreciated by employees, including technical staffs. By now most of the firms have fully utilised the premises they acquired so that further development can only take place with the erection of new buildings. At this point, the question of industrial expansion in relation to national park amenities may well arise.

Lastly, mention should be made of a small industry which, after several centuries of activity, is now almost defunct. In the yawning entrance to the Peak Cavern at Castleton, long flat terraces cut into the limestone indicate the scene of the ropemakers' craft. The ropewalks extend for nearly 300 ft. into the cave and on the uppermost platform are several pieces of the old hand-worked apparatus, still occasionally used. Until very recently there was a small but specialised output, using cotton and hemp fibres, of such interesting items as bell ropes, sash cords for windows, clothes lines and ropes for small boats.

REFERENCES

Report and Analysis of Survey. Peak District National Park Development Plan (1955) and (1966)

Harris, H. *Industrial Archaeology of the Peak District.* David and Charles (1971)

15. The Peak as a National Park

> Go you and adorn your country.
>
> *William Pitt, Earl of Chatham*

> Go you, my lord, and preserve it.
>
> *Lancelot (Capability) Brown*

In the account of the natural and human aspects of the Peak District which forms the subject of the preceding chapters, attention has constantly been drawn to the contrasts between the grit and shale country on the one hand and the limestone country on the other. This two-fold division, founded on geological considerations, affords the proper scientific basis for a description of the region and for a true understanding of its landscape. Only from a full recognition of the characteristics of these main divisions and of the variety of detail occurring within each of them can a real appreciation be gained as to what the National Park has to offer in amenity, recreation and scientific interest.

It is true that tradition has given rise to a division of the region into the High Peak and Low Peak, terms which still have some recognition though they really have no scientific significance. Broadly speaking, the High Peak refers to the higher northern portion and includes the gritstone moors as well as the limestone area approximately north of Lathkill Dale, while the Low Peak includes the rest of the limestone plateau, together with the flanking grits. The two terms have always been loosely used and the areas to which they relate have never been precisely defined, while the implied contrast between them as regards elevation is only partially true. There is, admittedly, some historical significance attaching to the two areas. For a long time after the Norman Conquest a large part of the feudal domain of the Peverils at Castleton covered the northern area; on the other hand, the territory of the Low Peak was not subject to any single overlord. Over much of this period a good deal of the High Peak was occupied by the royal hunting ground of the Peak Forest and cultivation, if not actually forbidden, was discouraged by penalties inflicted through the Forest Law. Eventually as the source of lead production, the whole limestone tract became known as the King's Field. Yet for the regulation of mining activities this was divided between the High Peak and Low Peak by the establishment of a Barmote court for each area.

At some stage the administrative area known as the High Peak Hundred came to be recognised, but the corresponding division embracing the Low Peak was called the Wirksworth Hundred. Today, long after the Hundreds have been superseded by modern local government divisions, High Peak remains the name of the parliamentary constituency covering north-west Derbyshire. Other than this, no real importance is any longer attached to either of these terms, and in many respects their use is more confusing than helpful. It is unquestionably more valuable to accept the two main types of scenery as the proper basis for a sub-division of the Peak District. The essential differences between the two landscapes, however greatly we may wish to amplify them in detail, can be very simply stated. In the case of the gritstone scenery the most impressive features are the steep, rugged 'edges' and the vast interminable moors beyond them, where a deeper solitude may be felt than almost anywhere in England. On the other hand, the most arresting scenery of the limestone country is in the deep verdant dales or even below ground in the dank, mysterious caves.

Although it was primarily on account of its scenic qualities that the Peak District became a national park, it is important to realise how truly varied are the opportunities which the area offers for the interest and satisfaction of the public. These opportunities, moreover, include much that stems from the human activity of the region. While the Peak District offers scope for many special forms of outdoor activity such as angling, grouse-shooting and rock-climbing, there is no doubt that the great majority of people find their enjoyment in traversing the hilly ground, observing the magnificent views and breathing the upland air. The good roads of today enable sightseers by car and coach to see much of the National Park in a single outing at an unhurried pace. Motor vehicles can reach most of the well-known viewpoints such as the famous Surprise View as seen from above Hathersage, which is so well-known to thousands of Sheffield citizens, the great sweep of country stretching eastwards from Axe Edge and the sudden and dramatic prospect from Monsal Head, near Little Longstone. Among many other accessible viewpoints, are the 'Cat and Fiddle' inn on the Buxton-Macclesfield road – from which an immense

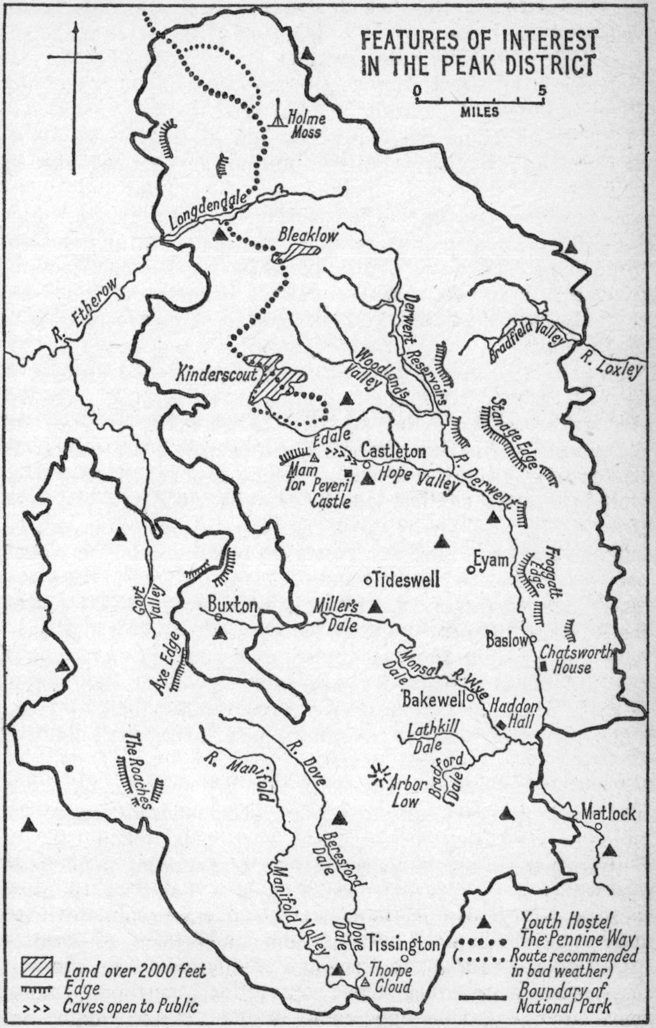

Fig. 14 Features of interest in the Peak District

spread of country, including the Goyt Valley, can be seen – and Alport Hill, near Wirksworth, although it lies a little outside the National Park, from which a vast panorama spreads southwards, giving in clear weather glimpses of the Wrekin and the Clee Hills some fifty miles away.

There is also little doubt, although there are no statistics to prove it, that away from the main roads, the largest category of persons seeking open-air recreation, is that of the hikers and ramblers. Such people, who journey on foot, whether as individuals or in groups, combine enjoyment of the countryside with healthy exercise, and to them, not unnaturally, the loftier and wilder areas such as Kinderscout, Mam Tor and Rushup Edge, the Roaches and Thorpe Cloud have a special attraction. Incidentally, to reach the summit of Kinderscout there are several approaches, the commonest and least troublesome being that by means of Jacob's Ladder from Edale and Barber Booth. This path is so named because it was first opened up by one Jacob Marshall. But that is not all, for the region is threaded with tracks and footpaths which lead to scores of places beyond the reach of wheeled traffic. The reward of the walker is immeasurably richer than that of the motorist or even the cyclist. Only by walking can many of the greatest delights of The Peak be experienced. Only by this means can the tops of the 'edges' and the tracks along their crests be reached, and only thus can Dovedale itself, or the beautiful Lathkill Dale and the many lesser dales be traversed. There can be surprises, too, for over the Roaches and nearby moors the wanderer on foot may once have caught sight of a wallaby. Such a startling discovery is not difficult to explain, however, for some years ago a few of these fascinating creatures escaped from Swythamley Park, since when they became acclimatised and for a long time bred successfully in secluded places.

Thus although a wider measure of public access is desirable and many more footpaths are needed, even under present conditions the Peak District is assuredly good walking country.

Among the projects designed to extend walking facilities in the National Park is the promotion of a long-distance route to be called the Pennine Way, which is to run northwards from The Peak along the moorland spine of northern England as far as the Scottish border. The idea of this continuous footpath was first suggested in 1935 by T. C. Stephenson, journalist and rambler, and was largely inspired by the analogy of the Pilgrim's Way and by the Appalachian Trail and the famous

John Muir Trail in the USA. The project quickly gained support, especially among the ramblers' organisations, but it also caught the imagination of many influential people, and in due course provision was made for the development of such long-distance paths under the National Parks Act of 1949. Official approval for the Pennine Way was granted in 1951. By using, wherever possible, existing paths and old half-forgotten tracks, this route is already complete except for two or three miles. It offers something of the appeal attaching to the Pilgrim's Way, with the added attraction of days of uninterrupted walking over high ground giving ever-changing views. The Pennine Way is now open and the track, which starts at Edale, runs for 250 miles to the Cheviots (Fig. 14).

It is interesting to note that the Board was responsible for setting up the first National Park youth hostel at Crowden in Longdendale close to the Pennine Way. It is operated by the YHA but also serves non-members.

THE GRITSTONE COUNTRY

Special activities for which the gritstone country offers opportunities are grouse-shooting, rock-climbing and winter sports. Grouse-shooting, while engaging only a small number of people for a limited period of the year, demands the use of an extensive area of moorland. The widespread occurrence of heather moor, on which grouse depend for the greater part of their food supply, enables these birds to be reared on some fifty individual 'moors' in the northern half of the National Park. Since the adoption of the breech-loading sports gun about a hundred years ago and the still more recent practice of driving to the butts, the annual rate of killing has increased, and this has necessitated more careful management of the moors in order to maintain their carrying capacity. This is done by improving the heather through rotational burning and by extending its area as the result of draining adjacent stretches of cotton-grass. While public access to the grouse moors is still restricted, complete exclusion seems to be no longer insisted upon. It should be noted, too, that, unlike some forms of recreation, grouse-shooting contributes to the local economy, for not only are rates chargeable to the local authority, but during the shooting season the driving of the birds provides employment for a considerable number of men. On the other hand, grouse-shooting is unprofitable in the commercial sense, the cost of each bird killed being nearly twice the amount received from its sale.

As far as rock-climbing is concerned, many of the gritstone edges afford suitable conditions for training as well as providing scope for the expert. Although the vertical faces are only of moderate height, the climbing presents varying degrees of difficulty. On the other hand, the limestone cliffs and scars are generally unsuitable owing to the danger of loosened rocks resulting from weathering along the joint planes. An exception is to be found in the Harborough Rocks, near Brassington, where the limestone is dolomitic and much tougher. Of the gritstone localities most frequented by rock-climbers, Laddow Rocks, near Longdendale in the north, Birchen Edge and Stanage Edge on the east and Windgather Rocks, the Roaches and Hen Cloud on the west should be mentioned, while the Black Rocks at Cromford are well-known among proficient climbers.

For many years past, whenever conditions have been favourable, local people have enjoyed ski-ing on slopes near Buxton and around Edale. On this account the latter has become known as Derbyshire's 'Little Switzerland'. Ski-ing and tobogganing have recently grown in popularity and, thanks to prolonged efforts on the part of the Peak Planning Board, a ski-run has been provided at Edale and was first brought into use during the snowfall of January, 1960. In time, as facilities improve, winter sports in The Peak are likely to attract increasing numbers of people.

One other activity associated with the configuration of the gritstone should be noted. At Hucklow Edge the effect of the height and westward exposure of the grit escarpment in promoting rising air currents (thermals) has made Camp Hill a favourable centre for the modern sport of gliding.

THE LIMESTONE COUNTRY

To pass from the gritstone to the limestone is largely to leave the moors for the dales and there is a corresponding contrast to be found in the forms of recreation. For the dales are the haunt of the angler, the clear streams being famous for their trout and grayling. The Dove, which Charles Cotton described as 'the Princess of rivers', continues to be a favourite resort as it was in the days of the squire of Beresford and his friend Isaak Walton, but today the Wye and the Derwent yield the best rainbow trout as well as the common brown trout. The fishing rights on these and other rivers, like the Lathkill and the Dane, are mostly privately owned, but a number of hotels hold rights over certain lengths of river, thus making angling less exclusive than it might otherwise be. In sharp contrast to the restful use of rod and line is the exploration of the limestone caves and underground passages, an activity which makes a strong appeal to both the adventurous and the scientifically minded. Many of the lesser known caves invite serious investigation, but this is a strenuous pursuit demanding both teamwork and individual skill. Several caving organisations, which are the equivalent of the pot-holing clubs of Yorkshire, make The Peak their main field of activity, and to illustrate the fact that there is still much to be learned from such exploration it is only recently that one of these parties discovered human skeletons of probable prehistoric date in one of the caves in the Longnor district.

In order to encourage outdoor interests among young people the Derbyshire County Council have established an Open Country Pursuits Centre. This admirable venture, the first of its kind in the country, was started in 1951 in order to give training in walking, climbing, ski-ing and caving, largely by week-end courses. Having its headquarters at White Hall, a country house near Buxton, the Centre is directed by a warden and a small permanent staff, assisted by voluntary helpers. For training purposes, Windgather Rocks as well as Birchen Edge, near Baslow, are used for rock-climbing, and the Giants' Hole Cavern, near Castleton, for teaching the techniques of caving.

As part of a long-settled country rich in tradition and past achievements, the Peak District has its share of historical features and many of these are additional sources of attraction to people visiting the region primarily for the enjoyment of its scenery. To some they are undoubtedly the main objects of interest, especially in the case of monuments of national importance. Several thousand people in the course of a year, for example, make their way to the great stone circle at Arbor Low, while even greater crowds visit the noble mansions of Chatsworth and Haddon Hall, as well as Lyme Hall on the western boundary of the Park near Stockport. In recent years over 200,000 people visited Chatsworth House and some 60,000 went to Haddon Hall annually. Several fine churches are also to be seen, such as those of Ashbourne, Bakewell and Tideswell, and of course there are the places, inevitably mentioned in guide books, which are connected with some particular incident of history. Although it is outside the scope of this book to deal with these, a typical example which might be mentioned is the much-visited Mompesson's Well at Eyam. Eyam was one of the worst afflicted villages in the country at the time of the plague, which spread from London towards the end of 1665. For over twelve months the stricken community maintained itself in voluntary isolation, thus preventing the scourge from affecting a wider area. Throughout this desperate period the rector, the Rev. William Mompesson and his wife, with a few staunch friends, displayed remarkable fortitude and calm in keeping control of the situation, although more than two-thirds of the 350 inhabitants died. Food was brought from other villages and left at outlying places, one of them being beside a little brook just to the north of the village. In remembrance of the rector's heroic struggle, this spot has ever since been known as Mompesson's Well. It is said that visitors today sometimes throw coins into the cleansing water to keep up the tradition of paying for the supplies brought by the neighbouring countryfolk at great risk to themselves.

At Eyam an annual commemoration service is held in the open-air on Plague Sunday, the last Sunday in August, for during the pestilence Mompesson conducted his services in the open in order to reduce the risk of villagers infecting one another. Reference to this event brings to mind the many other old customs which are still observed in the Peak District and which provide yet another source of interest to visitors. They also shed some light on the earlier conditions under which the local communities existed.

It is probable that some of the village well-dressing ceremonies rank among the most ancient of these customs, though some, apart from a few revivals, are entirely modern. At Tissington, the ceremony is held on Ascension Day and is said to date from the great drought of 1615, though there is nothing to prove that it did not originate far earlier, possibly in pagan times. This village has five wells and for the occasion each is decorated and each is visited in turn and blessed. Well-dressings have become more numerous in recent years, and in a recent summer the custom was observed in at least a dozen places. During June and July there were ceremonies at Ashford-in-the-Water, Buxton, Hope, Tideswell and Youlgreave, and in August at Barlow, Bradwell, Bonsall, Eyam, Stony Middleton and Wormhill.

At Castleton the rite of 'garlanding' is observed each year, when a 'king' decked with a garland, accompanied by a young woman (formerly a boy) dressed as his 'queen' in seventeenth-century costume, pass in procession through the village before the garland is hoisted on to the church tower. Garland Day is commemorated on Oakapple Day, May 29th, because the custom has for some reason or other, quite wrongly, become associated with the incident of Charles II and the oak tree; it is actually of much earlier origin. Far older, however, is the ceremony of rush-bearing at the tiny village of Wildboarclough in Macclesfield Forest, a relic of the times when church floors were strewn with either rushes or fresh straw. Very different from these survivals is the boisterous football game played at Ashbourne on Shrove Tuesday. This rough-and-tumble contest has been held for centuries. Scores of people still take part, those born in the upper half of the town playing for the 'up'ards' and those born in the lower half for the 'down'ards'; the goals are still the two mills on the Henmore Brook at either end of the town; the pitch is still about three miles long and of indefinite width except that play is no longer permitted in the main streets. And, as if to emphasise the differences between local customs, while Ashbourne is engaged in its furious Shrovetide battle, Winster marks the day with a pancake race along the village street.

Agricultural events also draw many visitors. The most important of these is the Bakewell Agricultural Show which, besides being the outstanding occasion for the farmers of The Peak, attracts competitors from all parts of the country. Of more than local interest, too, are the sheep-dog trials held at Longshaw, Hope and Dovedale, and the Chatsworth Horse Trials and Game Fair.

NATURE CONSERVANCY

This brief outline of the recreational pursuits to be found within the National Park, including the different items of cultural interest, serves to show how diversified are the opportunities which it offers for public enjoyment. There is, however, another important use of the Park still to be mentioned. The Peak as an area of distinctive landscape, in which Nature holds predominance in some parts and in others yields place to Man's interference, is a region of great significance for scientific investigation. It affords singular advantages for field studies, not only in the natural sciences (geology, botany and zoology), but in archaeology, social and economic history and geography.

In addition to excursions and field courses organised by numerous schools and colleges, no less than five universities which lie close to its margins, find it not only a valuable field laboratory but an area which poses innumerable problems for the research worker. The first National Park residential centre, supported by the Countryside Commission, was opened in 1972, at Losehill Hall near Castleton. This was recognised by the National Parks Act of 1949, under which many localities have been scheduled for preservation on the grounds of their special scientific importance. These are listed and briefly described in Appendix III. The National Parks Act also gave powers to the Nature Conservancy, which was set up in 1949, to acquire areas of scientific interest and to maintain them as nature reserves. So far the Nature Conservancy has no territory in the Peak District, but it is hoped that before long Dovedale Wood will pass into its hands, for this primitive ashwood on the west slope of Dovedale between Ilam and the Twelve Apostles is of major ecological importance.

On the other hand, the National Trust, a much older organisation, owns many sites in the region ranging in size from a few acres to several hundred acres. The protection of these areas has been secured not so much on scientific grounds, though this applies to a few, but on grounds of amenity for they mostly embrace stretches of either wild or particularly beautiful country, though they may simply consist of a building of either historical interest or outstanding architectural merit. While the National Trust properties are widely scattered throughout The Peak, there is not unexpectedly some concentration in particular districts such as the Dove and Manifold valleys, the Hope-Castleton-Edale district and the Upper Derwent. It will be appreciated that the first of these is a typical portion of the limestone country, the second includes both limestone and gritstone and the third entirely gritstone. In all, these properties cover more than 12,500 acres, while a further 3,500 acres, consisting of areas not owned by the Trust, are protected by restrictive covenants. In 1966 the Peak District was the first National Park in Europe to be awarded a diploma by the Council of Europe.

PROBLEMS OF THE NATIONAL PARK

In administering a national park, the Park Authority (in the case of The Peak, the Park Planning Board) has the twofold responsibility of ensuring the continuance of the existing economy on behalf of some 40,000 people and of promoting the functions of the Park as an area of public amenity and recreation. In connection with the latter, it should be noted that the Authority has powers to exercise control over new developments in the interests of the wider community. In The Peak itself the chief problems of management relate to the control of mineral working in respect of landscape preservation, the provision of accommodation for visitors (hostels, camping and caravan sites, car-parks and catering facilities) and, as in all national parks, the question of access and rights-of-way. The last-named problem is one of great importance but it is also one of the most difficult to deal with.

As far as access in general is concerned, the designation of a national park confers no further rights on the public than those previously enjoyed. Thus increased access to open country can only be obtained by further agreements made between land-owners and the Park Authority. Open country is defined in the National Parks Act (1949) as 'consisting wholly or predominantly of mountain, moor, heath, down, cliff or foreshore'. In the case of The Peak nearly half the total area falls within this definition, chiefly as mountain or moorland, consisting of the northern gritstone area and its extension along the eastern and western flanks of the limestone plateau. By far the greater part of this territory is not officially open to the public, although adequate access to parts of it is permitted by the owners concerned. Yet this is the country to which the greatest number of walkers and ramblers are attracted. It is the policy of the Park Planning Board, however, to negotiate wherever possible for the extension of public access. In 1955 an agreement was made covering the greater part of the Kinderscout Plateau and in 1958 another was concluded which freed the south-eastern portion of the neighbouring Bleaklow Plateau. As a result of these and other agreements thirty-seven square miles of the highest and wildest ground in The Peak have become open to walkers, while not far away to the east the Derwent Edge property of nearly 6,000 acres, owned by the National Trust, is also open to the public (Fig. 15). To ramblers these are substantial gains in the campaign for unfettered movement over the moorlands, and it is to be hoped that in due course there will be further advances.

Two special difficulties arise in relation to access which in themselves reflect some of the major characteristics of The

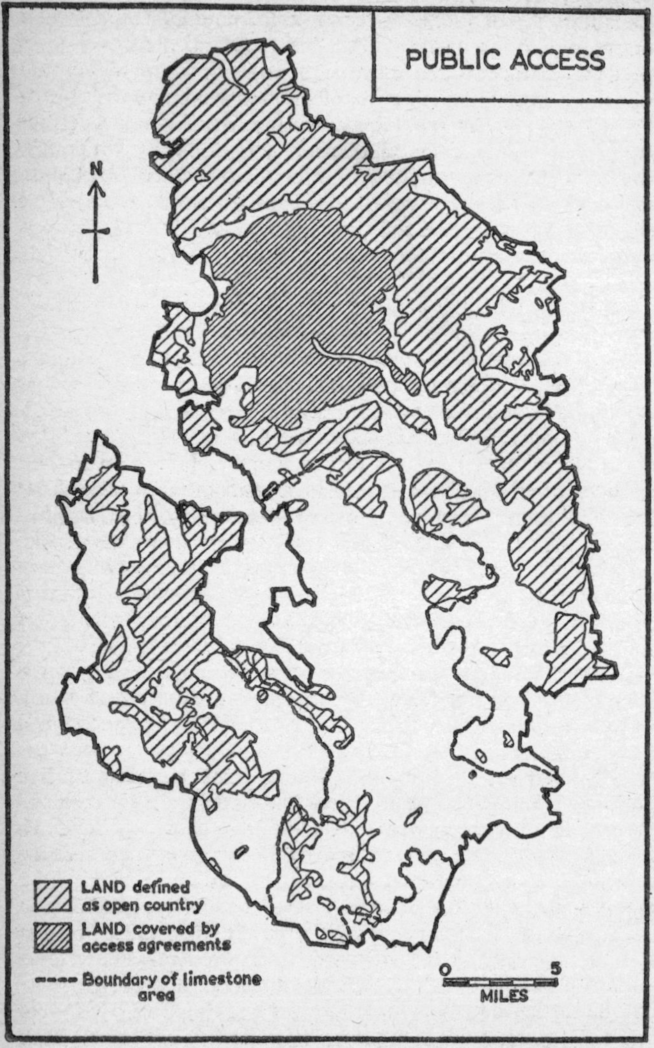

Fig. 15 Areas of public access in the Peak National Park

Peak. The first is the conflict of interests which exists between the use of the moorlands as water-gathering grounds and their availability to the public for recreation. The policy of the water supply authorities, leading to further and further restriction as the demand for water increases, though understandable and in some respects justifiable, is in direct opposition to the objects for which the National Park was created. The issue is a serious one and doubtless some compromise solution must eventually be found. The other difficulty concerns the risk of damage to cultivated land at the lower edge of the moors caused by people attempting to reach the open country above. This can be overcome, however, by the provision of new footpaths or the sign-posting of existing ones to serve as recognised approaches to the high ground.

On the related question of rights-of-way the Park Planning Board is confronted with a formidable task. The arrangements for confirming existing rights applicable to footpaths, bridleways and green tracks, and for the establishment of new footpaths, are extremely complicated and are only in part the responsibility of the Board. In general the maintenance and creation of footpaths rest with the local authorities, and the function of the Board is to advise on the general system. For this or other reasons it is widely held that little headway has been made; indeed the chief criticism of the progress so far achieved in national park development is directed to the disappointing results in the provision of footpaths. In The Peak it is feared that the situation instead of improving has deteriorated. In 1956 the Peak District and Northern Counties Footpaths Preservation Society reported that, over the entire area represented by the Society, more than 220 orders for the closure of footpaths had been approved and only five refused; nearly 650 orders for diversion had been accepted and only six refused, and only five new paths had been sanctioned. The report rightly stressed the fact that those who use existing footpaths can do much by their own efforts to save them. In fact the strongest defence of them is to use them. While it is the avowed policy of the Planning Board to do its utmost to defend the paths and to create new ones where they are required, it is clear that the position must be carefully watched by all who take any interest in the legitimate use of the Park for outdoor recreation.

From all that has been said in the foregoing pages it should now be reasonably clear that in the area known as the Peak District the interplay of natural environment and human activities has resulted in the promotion of a unique landscape, unique because in no other area either in this country or elsewhere, has the prolonged interaction of Nature and Man taken the same course or expressed itself in quite the same way. Man's activity alone, disregarding significant differences in the natural environment, has made the Peak District distinct even from the rest of the Pennines. In the course of time, moreover, further distinction will be imparted to the Peak landscape because of its role as a national park. This will follow, not from the mere fixing of a boundary defining the limits of the park, but from the fact that, within those limits, the area will be increasingly affected by the various measures enabling it to function as a national park. It has ceased to be an area significant only for its local life and its chance attraction for city-dwellers from the adjoining lowlands. It is true that the existing economy will be maintained but, with certain reservations such as the safeguarding of water supply which is a vital resource for a large population outside the Park, its amenities will be preserved and expanded to meet a national as well as a regional need.

In the meantime improved facilities are required for the people who live in the Park. Many of these communities are lacking in certain services such as secondary schools, clinics and fire services, and in social amenities such as village halls. Instead of piecemeal measures by which to make good these deficiencies, a careful scheme has been proposed by the Park Planning Board by means of which the bulk of such developments, including new housing, will be concentrated in selected villages, each of which will be able to serve its own area of countryside. These 'key' villages, as they are termed, will include Calver, Hartington, Hope, Parwich, Taddington, Tideswell, Winster and Youlgreave in Derbyshire, and Warslow and Waterfall-Water Houses in Staffordshire.

At this point we may well glance back and recall the inspiration given by the late Sir Patrick Geddes to the movement for improving the conditions of urban life and for the preservation of natural scenery. To Geddes, mountains and moorlands assumed the dual function of being an essential source of pure water and of providing scenery and pure air for public recreation. In his *Cities in Evolution*, first published in 1915, he declared:

> With this preservation of mountains and moorlands (i.e. for water supply) comes also the need of their access: a need

for health, bodily and mental together. For health without the joys of life – of which one prime one is assuredly this nature-access – is but dullness; and this we begin to know as a main way of preparation for insidious disease. With this, again, comes forestry: no mere tree-cropping but sylviculture, arboriculture too, and park-making at its greatest and best.

Here indeed is a prophetic image of what has been embodied in the national parks of today, with the main elements of its composition bearing a singular resemblance to those of The Peak.

National parks are to a large degree a reflection of the age in which we live, an age in which Man, through his triumphs over Nature, finds himself increasingly out of touch with her. They are in fact a timely device by which something of that intimacy may be restored. Man out of touch with Nature, relying only on himself, is an unworthy existence, as Wordsworth recognised when he wrote the lines:

and much it grieved my heart to think
What man has made of man.

As the first of the national parks in this country, The Peak not only serves this need but in one respect excels all others in so doing. For no other national park is so immediately accessible to such a large population. Within its liberal bounds thousands of people may enjoy the beauty of large unspoilt tracts of country, now rendered safe from encroachment by extensive building and the vulgarities associated with it.

National parks have an even deeper significance. Their value for outdoor recreation is obvious enough but it is not only a question of physical recreation; it is also a question of spiritual exercise and enjoyment and therefore not one of material values alone. As G. M. Trevelyan once reminded us, without vision the people perish and without sight and experience of the beauty of Nature the spiritual power of our nation will be endangered by atrophy. All the more important, therefore, is the location of the Peak District in the midst of the greatest concentration of industrial centres in the country and all the more valuable the legacy of its landscape, which ranges from the deeply-cut dales with their sparkling streams to the frowning heights of Kinderscout.

REFERENCES

The Peak District a National Park. Joint Committee for Peak District National Park of Council for Preservation of Rural England (1946)
Kirkham, N. *Derbyshire.* Elek (1947)
Porteous, C. *The Beauty and Mystery of Well-Dressing.* Pilgrim (1949)
Stamp, L. D. *Man and the Land.* Collins: New Naturalist (1955)
Report and Analysis of Survey. Peak District National Park Development Plan (1955) and (1966)
Nicholson, E. M. *Britain's Nature Reserves.* Country Life (1957)
Porteous, C. *Portrait of Peakland.* Robert Hale (1963)
Building in the Peak. The Peak Park Planning Board (1964)
Poucher, W. A. *The Peak and Pennines.* Constable (1966)
Oldham, K. *The Pennine Way.* Dalesman Publishing Co. 2nd ed. (1968)
Tarn, J. N. *The Peak District National Park: Its Architecture.* P.P.P.B. (1971)

Appendix I

Note on Bird Life in the Peak District

Various notes and reports on ornithological matters have appeared in the *Journal of the Derbyshire Archaeological and Natural History Society* and the *Derbyshire Countryside*, while in the former an annual ornithological record has been included for many years past. The most recent work on the Peak District appears in the publications of the Derbyshire Ornithological Society which issues a monthly bulletin and an annual report (from 1955 onwards) and has recently produced a survey of the birds of Derbyshire edited by F. C. Hollands, F.R.C.S. In reality most of the common English birds are found in the region, only a few species being comparatively rare, and even these occur in the rest of the Pennines and other upland areas. For this reason only a selected number of species is referred to in this note, mainly in order to illustrate particular points in the relationship between bird life and local habitat.

Within the Peak District altitude and vegetation exercise a marked control over the distribution of birds. As with other phenomena, both natural and cultural, a sharp contrast exists between the bird life of the limestone country and that of the gritstone moors. On the limestone, with its lower mean elevation, vegetation is richer and more varied than on the gritstone, so that most of the birds common to the surrounding lowlands are found. In the valleys especially, where trees and bushes abound, lowland passerine species such as the marsh tit *(Parus palustris)*, the long-tailed tit *(Aegithalos caudatus)*, the tree creeper *(Certhia familiaris)* and wheatear *(Oenanthe oenanthe)* are frequent, whereas on the high moors they are either scarce or absent altogether. Over much of the area, however, the limited tree cover and the lack of hedgerows bordering the fields restrict the number of lowland species. In the wooded dales, e.g. Dovedale, redstart *(Phoenicuris phoenicurus)*, tree pipit *(Anthus trivialis)* and wood warbler *(Phylloscopus sibilatrix)* are common.

It is in the gritstone country with its bleak moorlands and relatively impoverished vegetation that the bird life of The Peak is more distinctive, though there are no species unique to the region. Typical of these uplands are the plateaux of Kinderscout and Bleaklow, each with deeply dissected edges fretted with torrent-courses known as cloughs. On the lower marginal slopes, fields of rough pasture with *Juncus* in damp places provide the breeding zone of the lapwing *(Vanellus vanellus)*, snipe *(Gallinago gallinago)* and skylark *(Alauda arvensis)*, all of which evacuate the scene after the breeding season and are rarely encountered in winter. Rough pasture and occasional cultivated fields form regular stopping-places during the spring migration of large numbers of golden plover *(Charadrius apricarius)* and curlew *(Numenius arquata)*. Rocky outcrops on the steep upper slopes ('edges') are favoured by the ring ouzel *(Turdus torquatus)*.

Only scattered woods, chiefly of poor quality oak and birch, are found on the lower margins of the moorland tracts but even these promote a greater variety of birds, at least locally. For example, observations made at Shire Hill Wood east of Glossop, where the ground flora consists chiefly of *Deschampsia flexuosa* and *Pteridium* and the trees are openly spaced, show that redstart, tree pipit and yellow hammer *(Emberiza citrinella)* are common, while a few nightjars *(Caprimulgus europaeus)* and wood warblers breed nearby. In the larger valleys such as Edale, Woodlands Valley and Longdendale the same species, together with the little owl *(Athene noctua)* and sparrow-hawk *(Accipiter nisus)*, are also found, while the dipper *(Cinclus cinclus)* and common sandpiper *(Tringa hypoleucos)* are characteristic of the streams. The heather-clad slopes of such valleys provide nesting-places for small numbers of merlin *(Falco columbarius)*, a bird which is sadly persecuted by shepherds and farmers.

On the true moorland bird communities vary largely in relation to the nature of the ground, i.e. whether damp or well-drained, and the resultant vegetation. On the gentler well-drained slopes supporting Calluna, red grouse *(Lagopus lagopus)* and meadow pipit *(Anthus pratensis)* the chief host of the cuckoo *(Cuculus canorus)*, which is quite common, breed freely. The flatter surfaces carrying a mixture of *Calluna*, *Eriophorum* and *Nardus* are favoured by red grouse, golden plover and curlew in considerable numbers, while the meadow pipit also remains plentiful. Towards the windswept summit levels, approaching 2,000 feet in altitude, are the vast tracts of cotton-sedge *(Eriophorum vaginatum)* and peat moor which are persistently damp. They too are the haunts of the golden plover and curlew, while the redshank *(Tringa totanus)* is known to breed in places almost to 1,700 feet. A bird found almost exclusively on this high ground is the dunlin *(Calidris alpina)*, the typical habitat of which is *Eriophorum* interspersed with shallow pools and patches of bare peat, the nest itself being in the cotton-sedge.

With regard to aquatic species the Peak District is poorly represented, and even the large reservoirs of Longdendale and the Derwent are disappointing from the ornithological point of view. Apart from the moorhen *(Gallinula chloropus)* which breeds, mallard *(Anas platyrhynchos)* and teal *(Anas crecca)* visit some of the reservoirs regularly in small numbers, while the tufted duck *(Aythya fuligula)* appears less plentifully.

In general the Peak District has not received great attention from bird-watchers. Both its altitude and interior situation render it comparatively unsuitable for the observation of migrations and rare vagrants. Yet the bird life is as important an asset to the National Park as its other natural phenomena, the distinctive communities of the high moors providing an aspect of special interest. The species among those mentioned to be sought by the bird-watcher are the merlin, red grouse, golden plover, curlew, dunlin, common sandpiper, dipper and ring ouzel, all in their breeding quarters. These birds are typical of much of Highland Britain and the Peak District is the most accessible part of this zone for much of the Lowland. In this fact, it may be repeated, resides much of the naturalist's interest in the National Park.

Appendix II

Note on Fish Life in the Peak District Streams

The 'clear waters of Derbyshire', to use Isaak Walton's phrase, are famous among anglers. Since the time of *The Compleat Angler* (1676) the principal streams, the Derwent, Wye and Dove, have been especially noted for their trout and grayling. There are also ample records dating from the nineteenth century and the early twentieth to show that salmon from the Trent formerly reached these streams for spawning, but their numbers declined owing to various forms of interference, such as the growth of navigation on the Trent and the building of obstacles across the streams such as Darley Weir on the Derwent.

Within the uplands the Derbyshire streams have a relatively swift flow and are exceptionally pure. This is true, even of the Derwent below the great reservoirs, owing to the considerable volume of compensation water which is released from them into the stream. Only in their lowland courses, well beyond the southern limit of the National Park, do changes in physical conditions render the two main rivers, the Derwent and the Dove, more suitable for coarse fish than for trout and grayling. It is interesting to note that there is only one source of pollution from the industrial area to the east. This is the Blackwell Brook which flows westwards from the coalfield to join the Amber. As a result both the brook and the lower Amber are devoid of fish; however, the Amber, on entering the Derwent at Ambergate, has no harmful effects upon the coarse fish in that river. On the west side of the region the Churnet, which joins the Dove at Rocester, is also without fish, though the reasons for this are not yet understood.

For trout and grayling the chief rivers fall into three groups: the Derwent and its headstream tributaries, the Ashop and the Noe; the Wye and its tributaries, the Lathkill and Bradford; and the Dove and Manifold. While grayling are largely confined to these, trout are found in many other streams. Grayling, moreover, unlike trout, do not reach to the headwaters of the rivers, neither do they extend so far downstream. Their more limited distribution is illustrated in the case of the Derwent by the fact that they are found mainly between Cromford and Bamford Weir, whereas trout are abundant both above and below these points. Besides trout, the Derwent in its middle course between Ambergate and Matlock contains quantities of coarse fish such as perch, roach, chub and dace.

While the brown trout *(Salmo fario)* is the commonest variety found in the Peak District, the American rainbow trout *(Salmo irideus)* is present in the Wye above Haddon Hall, where it has found a suitable habitat and breeds successfully. Attempts to introduce another variety, the Loch Leven trout *(Salmo trutta levenensis)*, some fifty years ago, proved less successful despite careful protection.

In addition to the streams, a limited amount of fishing is permitted in some of the reservoirs belonging to the water supply authorities. The great Ladybower Reservoir is stocked with trout for this purpose, while other reservoirs situated at a lower altitude carry trout and coarse fish. Interesting results have been noted on the behaviour of the fish under these conditions. Thus R. G. Abercrombie has found that since the completion of the Ladybower scheme, trout in the Derwent below the dam have increased in size, while mayfly, formerly sparse, is now more abundant, perhaps as a consequence of the more regular flow of the river. In the reservoir itself the greatest concentration of fish occurs near the outlet of the Noe tunnel which brings water relatively rich in organic matter from the cultivated area of Edale.

Thus, whether in stream or reservoir, it is the common brown trout which is native to the whole of Britain and to much of Europe that provides the chief object of the anglers' sport in The Peak.

The principal angling clubs using the rivers of The Peak are the following: on the Derwent, the Chatsworth Fishery, the Darley Dale Fly Fishing Club, the Derwent Fly Fishing Club, the Ecclesbourne Fly Fishing Club, the Matlock Angling Club, the Matlock and Cromford Angling Association, the Sheffield Waltonians; on the Dove, the Ashbourne Fly Fishers Club, the Birdsgrove Fly Fishing Club; on the Wye, the Monsal Dale Fishery; on the Wye and Lathkill, the Duke of Rutland Estate Waters; and on the Amber, Derwent and Dove, the Sheffield Trout Anglers Association.

Appendix III

Sites of Special Scientific Interest

In accordance with Section 23 of the National Parks and Access to the Countryside Act 1949, all those areas or localities which have an outstanding interest on account of their geology, geomorphology, flora and fauna have been designated as Sites of Special Scientific Interest by the Nature Conservancy.

The Conservancy notifies such sites to the local planning authority, in this case the Peak Park Planning Board. The Board, in turn, is required to consult with the Conservancy before granting permission for development which might affect the scientific interest of these sites. (Article 11, Town and Country Planning General Development Order 1963.)

The designation of an area as an S.S.S.I., apart from denoting its importance and acting as a guide for planners, provides a means of informing the owner of the natural interest of his property should he be unaware of it. As a result an owner can avoid unwittingly destroying or damaging that interest, and can obtain the Conservancy's advice on how to protect it.[1]

Scheduled Sites within the Peak District National Park, March 1964

SITES OF GEOLOGICAL INTEREST

Bakestonedale The best section of Lower Coal Measure strata in this part of England.

Bradwell Dale One of the most accessible and best known exposures of the *Orionastrea* band.

Calton Hill An important example of varied volcanic rocks injected into the mountain limestone.

Calver Peak Quarry An exposure showing the rare occurrence of an unusual coral.

Castleton A tract of Carboniferous rocks of great and varied stratigraphic and physiographic interest. It includes important cave systems.

Chrome and Parkhouse Hills Fine examples of ancient limestone reefs projecting through surrounding shales. Famous for fossils.

Crossdale Head Quarry The most accessible exposure of the

[1] This list is based on the Schedule of Sites of Special Scientific Interest drawn up for the Peak District by the Nature Conservancy and is reproduced with the Conservancy's permission.

highly fossiliferous Hob's House Coral Band.

Dove Valley and Biggin Dale An area of exceptional geological and physiographical interest.

Edale The type succession of the Edale Shales.

Eldon Hole, Eldon Hill A famous pot-hole eroded in Carboniferous Limestone.

Grangemill Quarry Classic exposure of coarse conglomerate within a volcanic vent of Carboniferous age.

Harewood Grange Stream Section, Holymoorside Exposures from Crawshaw Sandstone to the Redmires Flags.

Hob's House A stack of Mountain Limestone including abundant and interesting corals.

Monsal Head The type locality of a rare coral. The highest coral bed of the D^2 succession in the area.

Station Quarry, Miller's Dale The type locality for the Station Quarry Beds and the Miller's Dale Beds of the Carboniferous Limestone.

Tideswell Quarry A bed of altered clay showing columnar structure due to the baking action of the overlying dolerite sill.

SITES OF BOTANICAL INTEREST

Bradford Dale A typical limestone valley. The varied woodland, grassland, stream and streamside habitats support a range of vegetation types including some nationally rare species.

Chee Tor and Chee Dale Limestone Tor and gorge with a great range of aspects and vegetational types. There is a particularly extensive stand of the very local Mountain Currant, *Ribes alpinum,* and Yew is abundant on the cliffs.

Clough Wood, Cowley Knoll, Hill Wood, Sabine Hey Wood, and Stanton Moor Edge A series of woodlands astride the limestone/Millstone Grit junction with interesting contrasts in composition and ground flora.

Coombs Dale A limestone Dale with relict ashwood and grassland with an outstanding flora including some very rare species.

Cressbrook Dale Botanically, one of the most outstanding Peak District limestone Dales with most of the major vegetation types and an extremely rich flora including many exceptionally rare species.

Dove Valley and Biggin Dale An area of exceptional botanical and ornithological interest. The famous relict ashwood on the west wall of Dovedale is of great ecological interest.

Goyt Valley One of the best areas on the west side of the National Park showing relict oakwood-land in the valley and the transition from valley woodland to moorland. A wide range of upland, woodland and water birds inhabit the area, and the insects, particularly Microlepidoptera, are of special interest.

Hallam Moors A characteristic part of the Millstone Grit escarpment overlooking the Derwent Valley. The bog to the east of the escarpment is one of the few high altitude peat areas to have escaped drainage or erosion.

Ible and Middleton Woods The greater part is covered by semi-natural ash woodland including some *Tilia cordata* together with calcareous scrub, limestone grassland and lead spoil vegetation. Several rare plant species occur within the area.

Kinder and Bleaklow An area of outstanding importance for its upland acidophilous vegetation showing certain features not encountered elsewhere in the British Isles.

Ladybower Tor An area of moorland which is one of the two most southerly stations for a plant which is largely restricted to Scotland.

Lathkill Dale A deep limestone valley through which runs one of the only limestone streams in the Peak District not arising from an allogenic source. There are extensive relict ash woodlands and steep limestone grassland slopes on a variety of soils.

Leash Fen A small marshy part of an extensive cotton grass moorland in which occur two semi-aquatic plants which are rare in the district.

Leek Moors An extensive area of botanical and ornithological interest. Some of the undrained mosses and bogs support a rich flora including rare species.

Longdale and Gratton Dale Dry limestone dales with a variety of soils and slopes of all aspects varying from horizontal to 40 degrees. The area is of considerable phytogeographical importance.

Mill Clough, Derwent A moorland clough of phytogeographical interest as locality for a plant at the north-eastern limit of its distributional range.

Miller's Dale A limestone Dale in which there is evidence of rapid successional change from limestone grassland to scrub and climax woodland.

Monk's Dale An outstanding limestone Dale including representatives of major vegetation types. The flora is extremely rich and contains many exceptionally rare species.

Moss Carr A marsh with well-developed willow-carr and valuable for its orchid flora; also of ornithological interest.

Padley Woods The best example in the National Park of an oak woodland on acid soils, and noted for its variety of mosses. The range of birds occurring and breeding is particularly extensive.

Parwich Moor A small area of heather moor of pedological interest. Mounds of mineral soil exhibit evidence of sorting and rearrangement by frost and wind action.

Ramsley Moor A strip of marshy vegetation alongside a moorland stream which contains an unusually wide range of rush species and plants uncommon to the Peak District.

Taddington Wood An ashwood on steep slopes with a typical shrub layer and ground flora. Several very rare species occur here.

Topley Pike and Deepdale A limestone dale of prime phytogeographical importance containing large populations of at least three species which are rare or restricted in the British Isles.

Additional Sites scheduled in 1972

SITES OF GEOLOGICAL INTEREST

Alport Castles One of the largest and most complex landslip features in Great Britain.

Alport Tufa Exposure The best tufa exposure in the Peak District.

Ashford Black Marble Quarry The only place in England where black marble has been quarried.

Bretton Clough An extensive upland gritstone valley with a large and complex landslipped area.

Hamps and Manifold Valleys Ecton Hill shows a unique mineralisation and structure and is noted for its copper mines which used to be the richest in Europe.

Lee Farm Quarry Exposures in the D1 zone of the Carboniferous Limestones showing several large and important coral colonies, which have a bearing on limestone sedimentation and coral growth theories.

Low Moor Silica Sand Pit This site consists of three pits which represent subsidence hollows in the Carboniferous Limestone which now have a sand and clay fill of supposed mid-Tertiary age.

Monsal Head Railway Cuttings These cuttings provide an almost complete section of the Upper Dibunophyllum zone of the Carboniferous Limestone.

Shining Cliff The type locality for the Shining Cliff Beds, and a particularly useful section for D2 zone Carboniferous Limestone stratigraphy.

Sparrow Pit An important locality in the D2 zone of the Carboniferous Limestone.

Stony Middleton Crag An exposure of a coral bed in the Carboniferous Limestone which is the type locality for one coral species.

Wyns Tor A good example of a Tor in dolomitised limestone.

SITES OF BOTANICAL INTEREST

Abney Clough A small, damp gritstone valley which has an exceptional range of fern species including two which are near the edge of their range and which are extremely local in the Peak District.

Bretton Clough An extensive upland gritstone valley with varied plant associations.

Hinkley Wood A valley woodland on Carboniferous Limestone noted for the unusual abundance of Small-leaved Lime *(Tilia cordata)*.

Houndkirk Moor A wet flush in acid moorland which is the only locality in the Peak District for a rare plant.

Oliver Hill A moorland site of phytogeographical interest. One of the most southerly stations in the British Isles for Cloudberry *(Rubus chamaemorus)*.

Index

Fontana Introduction to Modern Economics

General Editor: C. D. Harbury

Each of the seven books in the series introduces the reader to a major area or aspect of modern economics. Each stands on its own, but all fit together to form an introductory course which covers most A-Level and first year university syllabuses, and those of most professional bodies.

Already published

Income, Spending and the Price Level A. G. Ford **75p**

An Introduction to Economic Behaviour C. D. Harbury **60p**

International Trade and the Balance of Payments
H. Katrak **50p**

Mathematics for Modern Economics R. Morley **£1**

Private and Public Finance G. H. Peters **60p**

Britain and the World Economy, 1919-1970
L. J. Williams **50p**

For publication 1973

Economics of the Market G. Hewitt